Lecture Notes in Computer Science   15976

Founding Editors

Gerhard Goos
Juris Hartmanis

Editorial Board Members

Elisa Bertino, *Purdue University, West Lafayette, IN, USA*
Wen Gao, *Peking University, Beijing, China*
Bernhard Steffen ⓘ, *TU Dortmund University, Dortmund, Germany*
Moti Yung ⓘ, *Columbia University, New York, NY, USA*

The series Lecture Notes in Computer Science (LNCS), including its subseries Lecture Notes in Artificial Intelligence (LNAI) and Lecture Notes in Bioinformatics (LNBI), has established itself as a medium for the publication of new developments in computer science and information technology research, teaching, and education.

LNCS enjoys close cooperation with the computer science R & D community, the series counts many renowned academics among its volume editors and paper authors, and collaborates with prestigious societies. Its mission is to serve this international community by providing an invaluable service, mainly focused on the publication of conference and workshop proceedings and postproceedings. LNCS commenced publication in 1973.

Esther Puyol-Antón · Enzo Ferrante ·
Aasa Feragen · Andrew King ·
Veronika Cheplygina ·
Melani Ganz-Benjaminsen · Ben Glocker ·
Eike Petersen · Heisook Lee
Editors

Fairness of AI in Medical Imaging

Third International Workshop, FAIMI 2025
Held in Conjunction with MICCAI 2025
Daejeon, South Korea, September 23, 2025
Proceedings

Editors
Esther Puyol-Antón ⓘ
HeartFlow and King's College London
London, UK

Enzo Ferrante ⓘ
National University of the Litoral
Santa Fe, Argentina

Aasa Feragen ⓘ
Technical University of Denmark
Kgs Lyngby, Denmark

Andrew King ⓘ
King's College London
London, UK

Veronika Cheplygina ⓘ
University of Copenhagen
Copenhagen, Denmark

Melani Ganz-Benjaminsen ⓘ
University of Copenhagen
Copenhagen, Denmark

Ben Glocker ⓘ
Imperial College London
London, UK

Eike Petersen ⓘ
Technical University of Denmark
Copenhagen, Denmark

Heisook Lee
Korea Center for Gendered Innovations for
Science and Technology Research
Seoul, Gangnam-gu, Korea (Republic of)

ISSN 0302-9743 ISSN 1611-3349 (electronic)
Lecture Notes in Computer Science
ISBN 978-3-032-05869-0 ISBN 978-3-032-05870-6 (eBook)
https://doi.org/10.1007/978-3-032-05870-6

© The Editor(s) (if applicable) and The Author(s), under exclusive license
to Springer Nature Switzerland AG 2026

This work is subject to copyright. All rights are solely and exclusively licensed by the Publisher, whether the whole or part of the material is concerned, specifically the rights of translation, reprinting, reuse of illustrations, recitation, broadcasting, reproduction on microfilms or in any other physical way, and transmission or information storage and retrieval, electronic adaptation, computer software, or by similar or dissimilar methodology now known or hereafter developed.
The use of general descriptive names, registered names, trademarks, service marks, etc. in this publication does not imply, even in the absence of a specific statement, that such names are exempt from the relevant protective laws and regulations and therefore free for general use.
The publisher, the authors and the editors are safe to assume that the advice and information in this book are believed to be true and accurate at the date of publication. Neither the publisher nor the authors or the editors give a warranty, expressed or implied, with respect to the material contained herein or for any errors or omissions that may have been made. The publisher remains neutral with regard to jurisdictional claims in published maps and institutional affiliations.

This Springer imprint is published by the registered company Springer Nature Switzerland AG
The registered company address is: Gewerbestrasse 11, 6330 Cham, Switzerland

If disposing of this product, please recycle the paper.

Preface

In recent years, research into fairness in machine learning has highlighted the significant risks associated with biased systems across various applications. Numerous studies have demonstrated that machine learning systems can exhibit biases related to demographic attributes such as gender, ethnicity, age, and geographical location, leading to unequal treatment of disadvantaged or underrepresented subpopulations. While fairness in machine learning has been extensively investigated in decision-making contexts such as job hiring, credit scoring, and criminal justice, it is only recently that researchers have begun to explore and address bias in medical image computing (MIC) and computer-assisted interventions (CAI).

To further this important discussion, the *3rd International Workshop on Fairness of AI in Medical Imaging (FAIMI 2025)* was held in Daejeon, South Korea, on September 23, 2025. Organized in conjunction with the International Conference on Medical Image Computing and Computer Assisted Intervention (MICCAI 2025), the workshop aimed not only to raise awareness about potential fairness issues in machine learning within the context of biomedical image analysis, but also to bring together researchers from the MIC, CAI, machine learning, and fairness communities to discuss bias assessment and mitigation strategies.

The workshop featured four key sessions: (1) two keynote presentations from expert speakers; (2) oral presentations by the authors of selected papers; (3) poster presentations; and (4) a round table discussion. All accepted papers were presented as posters, and attendees and program chairs jointly voted to determine the recipient of the Best Paper Awards.

The peer-reviewed papers were selected using the CMT system through a double-blind review process. Reviewers were recruited through an open call via the FAIMI newsletter, resulting in a large and diverse pool of 34 reviewers from 27 different institutions across 4 continents. This pool ranged from first-time reviewers to those with over five years of reviewing experience. To ensure a fair assessment of all papers while simultaneously providing a valuable learning experience for the diverse group of new reviewers, all submissions received three independent reviews. Assignments balanced experienced and new reviewers for each paper, with careful consideration given to potential conflicts of interest and recent collaborations among peers. The reviews were overseen by the organizing committee and culminated in a meta-review that informed the final paper selection. Out of 29 submitted papers, 21 were accepted for publication, and the top six were invited to deliver oral presentations.

This year, the Korea Center for Gendered Innovations (GISTeR) generously sponsored and presented the awards for the two top papers and the best poster.

We would like to express our profound gratitude to our program committee members, authors, and attendees, whose contributions made FAIMI 2025 a resounding success. We also extend our thanks to the FAIMI Advisory Panel for their invaluable advice and guidance, and to the Activity Committee for their tireless work in facilitating the practical aspects of workshop organization.

September 2025

Esther Puyol-Antón
Aasa Feragen
Andrew King
Enzo Ferrante
Veronika Cheplygina
Melanie Ganz-Benjaminsen
Ben Glocker
Eike Petersen
Heisook Lee

Organization

Program and Organizing Committee

Veronika Cheplygina	IT University Copenhagen, Denmark
Aasa Feragen	Technical University of Denmark, Denmark
Enzo Ferrante	CONICET, Universidad Nacional del Litoral, Argentina
Melanie Ganz-Benjaminsen	University of Copenhagen, Rigshospitalet, Denmark
Ben Glocker	Imperial College London, UK
Andrew King	King's College London, UK
Eike Petersen	Technical University of Denmark, Denmark
Esther Puyol-Antón	HeartFlow and King's College London, UK
Heisook Lee	GISTeR, South Korea

Advisory Panel

Judy Gichoya	Emory University, USA
Kanwal Bhatia	Aival, UK
Tal Arbel	McGill University, Canada
Bishesh Khanal	NAAMII, Nepal
Ira Ktena	Deep Mind, UK
Sanmi Koyejo	Stanford University, USA
Karim Lekadir	Universitat de Barcelona, Spain
Mercy Asiedu	Google Research, USA

Activity Committee

Tareen Dawood	King's College London, UK
Nina Weng	Technical University of Denmark, Denmark
Tiarna Lee	King's College London, UK
Dewinda Julianensi Rumala	Institut Teknologi Sepuluh Nopember, Indonesia
Emma Stanley	University of Calgary, Canada
Akshit Achara	King's College London, UK

Program Committee

Annika Reinke	German Cancer Research Center, Germany
Melat Ilmeya	Assam Science and Technology University, India
Théo Sourget	CentraleSupélec, France
Estanislao Claucich	Universidad Nacional del Litoral, Argentina
Melanie Ganz-Benjaminsen	University of Copenhagen, Rigshospitalet, Denmark
Dimitri Kessler	University of Barcelona, Spain
Gökhan Özbulak	École Polytechnique Fédérale de Lausanne, Switzerland
Maria A. Zuluaga	EURECOM, France
Oscar Jimenez del Toro	Idiap Research Institute, Switzerland
Kate Cevora	Imperial College London, UK
Raghav Mehta	Imperial College London, UK
Tian Xia	Imperial College London, UK
Amelia Jiménez-Sánchez	IT University of Copenhagen, Denmark
Veronika Cheplygina	IT University Copenhagen, Denmark
Fardin Afdideg	Karolinska Institutet, Sweden
Tiarna Lee	King's College London, UK
Tareen Dawood	King's College London, UK
Taha Emre	Medical University of Vienna, Austria
Cosmin Bercea	Technical University of Munich, Germany
Aasa Feragen	Technical University of Denmark, Denmark
Nina Weng	Technical University of Denmark, Denmark
Dimitri Kessler	University of Barcelona, Spain
Didem Stark	Bernstein Center for Computational Neuroscience Berlin, Germany
Tanya Akumu	University of Barcelona, Spain
Vien N. Dang	University of Barcelona, Spain
Emma Stanley	University of Calgary, Canada
Mariana Bento	University of Calgary, Canada
Matthias Wilms	University of Calgary, Canada
Raissa Souza	University of Calgary, Canada
Yuning Du	University of Edinburgh, UK
Anissa Alloula	University of Oxford, UK
Zikang Xu	University of Science and Technology of China, China
Xiaohao Cai	University of Southampton, UK

Contents

LTCXNet: Tackling Long-Tailed Multi-label Classification and Racial
Bias in Chest X-Ray Analysis 1
   Chin-Wei Huang, Chi-Yu Chen, Mu-Yi Shen, Kuan-Chang Shih,
   Shih-Chih Lin, and Po-Chih Kuo

Fairness and Robustness of CLIP-Based Models for Chest X-Rays 11
   Théo Sourget, David Restrepo, Céline Hudelot, Enzo Ferrante,
   Stergios Christodoulidis, and Maria Vakalopoulou

ShortCXR: Benchmarking Self-supervised Learning for Shortcut
Mitigation in Chest X-Ray Diagnostics 22
   You-Qi Chang-Liao and Po-Chih Kuo

How Fair are Foundation Models? Exploring the Role of Covariate Bias
in Histopathology 32
   Abubakr Shafique, Amanda Dy, Xiaoli Qin, Najd Alshamlan,
   Susan J. Done, Dimitrios Androutsos, and April Khademi

The Cervix in Context: Bias Assessment in Preterm Birth Prediction 43
   Joris Fournel, Paraskevas Pegios, Emilie Pi Fogtmann Sejer,
   Martin Tolsgaard, and Aasa Feragen

Identifying Gender-Specific Visual Bias Signals in Skin Lesion
Classification 53
   Heejae Lee, Sejung Yang, Yuseong Chu, and Byungho Oh

Fairness-Aware Data Augmentation for Cardiac MRI Using
Text-Conditioned Diffusion Models 63
   Grzegorz Skorupko, Richard Osuala, Zuzanna Szafranowska,
   Kaisar Kushibar, Vien Ngoc Dang, Nay Aung, Steffen E. Petersen,
   Karim Lekadir, and Polyxeni Gkontra

Exploring the Interplay of Label Bias with Subgroup Size and Separability:
A Case Study in Mammographic Density Classification 74
   Emma A. M. Stanley, Raghav Mehta, Mélanie Roschewitz,
   Nils D. Forkert, and Ben Glocker

Does a Rising Tide Lift All Boats? Bias Mitigation for AI-Based CMR
Segmentation 84
  Tiarna Lee, Esther Puyol-Antón, Bram Ruijsink, Miaojing Shi,
  and Andrew P. King

MIMM-X: Disentangeling Spurious Correlations for Medical Image
Analysis 94
  Louisa Fay, Hajer Reguigui, Bin Yang, Sergios Gatidis,
  and Thomas Küstner

Predicting Patient Self-reported Race From Skin Histological Images
with Deep Learning 104
  Shengjia Chen, Ruchika Verma, Kevin Clare, Jannes Jegminat,
  Eugenia Alleva, Kuan-lin Huang, Brandon Veremis, Thomas Fuchs,
  and Gabriele Campanella

Robustness and Sex Differences in Skin Cancer Detection: Logistic
Regression vs CNNs 115
  Nikolette Pedersen, Regitze Sydendal, Andreas Wulff, Ralf Raumanns,
  Eike Petersen, and Veronika Cheplygina

Sex-Based Bias Inherent in the Dice Similarity Coefficient: A Model
Independent Analysis for Multiple Anatomical Structures 125
  Hartmut Häntze, Myrthe Buser, Alessa Hering, Lisa C. Adams,
  and Keno K. Bressem

The Impact of Skin Tone Label Granularity on the Performance
and Fairness of AI Based Dermatology Image Classification Models 135
  Partha Shah, Durva Sankhe, Maariyah Rashid, Zakaa Khaled,
  Esther Puyol-Antón, Tiarna Lee, Maram Alqarni, Sweta Rai,
  and Andrew P. King

Causal Representation Learning with Observational Grouping for CXR
Classification 145
  Rajat Rasal, Avinash Kori, and Ben Glocker

Invisible Attributes, Visible Biases: Exploring Demographic Shortcuts
in MRI-Based Alzheimer's Disease Classification 156
  Akshit Achara, Esther Puyol Anton, Alexander Hammers,
  Andrew P. King, and for the Alzheimers Disease Neuroimaging Initiative

Fair Dermatological Disease Diagnosis Through Auto-weighted Federated
Learning and Performance-Aware Personalization 167
  Gelei Xu, Yawen Wu, Zhenge Jia, Jingtong Hu, and Yiyu Shi

Assessing Annotator and Clinician Biases in an Open-Source-Based Tool
Used to Generate Head CT Segmentations for Deep Learning Training 177
    Artur Paulo, Pedro Vinicius Silva, Tayran Mila Mendes Olegario,
    Paula Bresciani de Andrade, Klaus Schumacher,
    Rafael Maffei Loureiro, Joselisa Peres Queiroz de Paiva, Raissa Souza,
    and Bruna Garbes Gonçalves Pinto

meval: A Statistical Toolbox for Fine-Grained Model Performance Analysis ... 187
    Dishantkumar Sutariya and Eike Petersen

Revisiting the Evaluation Bias Introduced by Frame Sampling Strategies
in Surgical Video Segmentation Using SAM2 198
    Utku Ozbulak, Seyed Amir Mousavi, Francesca Tozzi, Niki Rashidian,
    Wouter Willaert, Wesley De Neve, and Joris Vankerschaver

Disentanglement and Assessment of Shortcuts in Ophthalmological
Retinal Imaging Exams 208
    Leonor Fernandes, Tiago Gonçalves, João Matos, Luis Nakayama,
    and Jaime S. Cardoso

Author Index 219

LTCXNet: Tackling Long-Tailed Multi-label Classification and Racial Bias in Chest X-Ray Analysis

Chin-Wei Huang[1], Chi-Yu Chen[2], Mu-Yi Shen[1], Kuan-Chang Shih[1], Shih-Chih Lin[1], and Po-Chih Kuo[1(✉)]

[1] National Tsing Hua University, Hsinchu, Taiwan
kuopc@cs.nthu.edu.tw
[2] National Taiwan University Hospital, Taipei, Taiwan

Abstract. Chest X-ray (CXR) classification faces challenges from long-tailed, multi-label data distributions and demographic biases in medical AI systems. To address these, we present LTCXNet a framework combining ConvNeXt, ML-Decoder, and multi-branch learning evaluated on Pruned MIMIC-CXR-LT dataset curated for long-tail scenarios. The model achieves large performance gains especially in rare classes, with 79% and 48% improvements in detecting Pneumoperitoneum and Pneumomediastinum respectively. We introduce "mAUCr" fairness metric to quantify racial group performance disparities, demonstrating LTCXNet's superior fairness in tail class subgroups compared to existing long-tail methods. This work advances medical imaging analysis by addressing both class imbalance and demographic bias through novel architectural integration and evaluation metrics. Our code is available on code.

Keywords: Long-tail · Multi-label · CXR · Class imbalance

1 Introduction

Deep learning Chest X-ray (CXR) model is one of the most common applications for medical AI. However, its real-world utilization faces multiple hurdles, including long-tailed distribution, multi-label classification [12,13], and racial bias [11]. Long-tailed distribution, characterized by skewed disease frequency, biases predictive models away from the detection of less frequent but critical diseases [2,17,19]. Accurately identifying these diseases is paramount, particularly when many of them pose severe health risks [23].

The complexity of medical image prediction is compounded by the multi-label nature of CXRs, where a single image can exhibit multiple diseases. This setup demands precise disease recognition and necessitates classifiers explicitly designed for multi-label tasks [18,25]. Meanwhile, the long-tailed issue poses another critical challenge: tail classes contain minimal samples and are thus

---

C.-W. Huang and C.-Y. Chen—Equal contribution.

highly vulnerable to even slight demographic biases, which can lead to disproportionately skewed results between demographic groups. Consequently, it is imperative to evaluate how a method's performance varies across demographic attributes and ensure equitable predictions in healthcare AI models [28,31].

Therefore, based on previous work [15] that combined ConvNext [21] and multi-branch learning (MB) [15], we introduced LTCXNet—which employs ML-Decoder (MLD) [25] as the classification head for CXR image prediction. ConvNeXt's transformer-inspired architecture offers enhanced performance over conventional convolutional neural network (CNN) models. MLD, a transformer-based classification head, excels in multi-label classification and reduces computational load. Multi-branch learning merges insights from models trained on different class subsets for more accurate and dependable predictions. We assessed different existing approaches using the Pruned MIMIC-CXR-LT dataset [12] and focused on their performance as well as demographic fairness. Our finding demonstrates LTCXNet's superior performance over current methods in long-tailed, multi-label classification, and highlights its fair rare disease prediction in different racial groups.

2 Related Work

Methods tackling imbalanced datasets can be categorized into four groups: decoupled representation learning, loss-modifying, dataset sampling, and backbone pre-training via self-supervised methods.

Decoupled representation learning [17,27] leveraged a two-stage method. In the first stage, image representation was learned with instance-balanced sampling. In the second stage, the classification head (i.e., the final layers of the model) was retrained on a class-balanced dataset. This approach is based on the assumption that earlier layers primarily focus on feature extraction, whereas the last few layers are more susceptible to the skewed distribution of the training data. On the other hand, the loss-modifying approach [8,20] gives different weights based on group sizes or their learning difficulty. Samples that are hard to learn are assigned higher weights to counteract the dominance of the majority class. Similarly, dataset sampling follows the similar principle but modifies the dataset distribution instead of the loss function. For example, ROS (Random Oversampling) [29] and ML-ROS [5] increases the frequency of minority class samples to enhance their influence during training, thereby improving worst-group performance. Lastly, backbone pre-training involves self-supervised learning (SSL) methods to develop a robust feature extractor. This line of methods includes SimCLR [6], DINO [3], I-JEPA [1], etc.

3 Method

3.1 Overview

Figure 1 describes our model architecture. We trained three distinct models, each focusing on a specific subset of labels: 'Head', 'Tail', and 'All'. The detailed

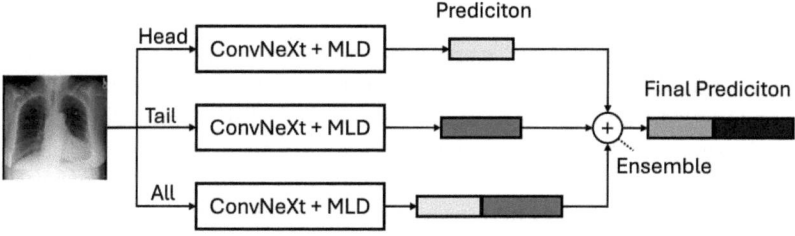

Fig. 1. Overview of the proposed method: Input images processed by three branches (Head, Tail and All). The predictions from the Head and Tail branches are first concatenated and then averaged with All branch. The length of the prediction block corresponds to the number of output classes.

division is described in Sect. 3.4. Each model consisted of ConvNeXt, positional encoding [30], and MLD. The final prediction was obtained by averaging the output probabilities from the 'Head', 'Tail', and 'All' models.

3.2 Dataset

We utilized Pruned MIMIC-CXR-LT [12], which comprises 257,018 frontal CXRs, each annotated with 19 labels that indicates the presence of the disease. This dataset showcases the multi-label, long-tailed feature of real-world medical imaging, where a few common findings are followed by many rarer conditions.

In our study, the dataset was divided into train, validation, and test sets, containing 182,380, 20,360, and 54,268 images, respectively. To ensure robust evaluation, we generated five subsampled test sets, each containing 80% of the original test set, using five different random seeds. Images were resized to 256 × 256 pixels and subjected to a series of data augmentation, including rotation, padding, brightness adjustment, Gaussian blurring, contrast manipulation, and posterization for preprocessing. Figure 2 illustrates the long-tailed nature of the dataset, with the most common class having 104,364 samples compared to just 553 samples in the least common class.

3.3 ConvNeXt with MLD

ConvNeXt [21] blends the robust feature extraction of CNNs with the contextual global comprehension of attention models. We chose it as our main backbone through an extensive architecture search across ResNet, DenseNet and Transformer models. In multi-label classification, transformer-decoder classification head proves beneficial but fails in scalability for its quadratic computation requirement. Meanwhile, MLD [25] offers a viable alternative by removing self-attention mechanisms and adopting group-decoding to improve both efficiency and scalability. Therefore, we selected MLD as our classification head.

3.4 Multi-branch Learning

In our study, 'Head' and 'Tail' were subsampled from the dataset, while 'All' represented the entire dataset. This division followed prior research [15], where 'Head' included the nine most prevalent classes, and 'Tail' contained the remaining ten classes and the 'Support device' category. 'All' division comprised all classes. Note that the 'Support device' category was included in both the 'Head' and 'Tail' divisions due to its high prevalence in the dataset. Excluding this category from the 'Tail' division would have resulted in insufficient training samples for model learning. To integrate predictions from different branches, the output probabilities of 'Head', 'Tail', and 'All' models were averaged to generate the final prediction.

3.5 Evaluation Metrics

We evaluated model performance using mean Average Precision (mAP) and macro F1 score (mF1) [10], which both treat each class equally and are suitable for imbalanced datasets. mAP averages the area under the precision-recall curve (AP) across all classes, while mF1 averages F1 scores calculated at optimal thresholds determined from the validation set. We determined the threshold for each class in F1 score by searching in the validation set. To quantify group fairness, traditional fairness metrics, such as predictive equality or equality of opportunity [4] are sensitive to threshold choices, so we introduced a new metric, **mAUCr**, which assesses the consistency of the area under the ROC curve (AUROC) across demographic groups.

In our analysis, we chose race as our primary target for analyzing demographic disparity since many disease classes exhibit strong correlation with other demographic attributes such as age or sex in their etiology. For example, "Calcification of the Aorta" is highly correlated with elderly [16]. Our race attribute contains five different kinds of races, including White, Black, Hispanic, Asian, and Other. The mAUCr is computed as:

$$mAUCratio = \frac{\min_D(\frac{\sum_i(AUC_{i,d})}{|I|})}{\max_D(\frac{\sum_i(AUC_{i,d})}{|I|})} \quad (1)$$

where $i \in I$ is a disease label, $|I|$ stands for the number of disease classes, $d \in D$ represents a demographic group, and $AUC_{i,d}$ symbolizes AUC for demographic d and disease label i. Practically, "Pneumoperitoneum" was ignored in the mAUCr calculation due to its scarcity in certain racial groups in testing datasets.

3.6 Implementation Details

Implemented in PyTorch [22], our model utilized ConvNeXt-small [21], pre-trained on ImageNet [9]. Binary cross-entropy loss was utilized with Lion [7] optimizer at a learning rate of 6×10^{-6} and weight decay of 5×10^{-5}. The batch size was 32. Each experimental result is the average result on the five resampled test sets unless mentioned.

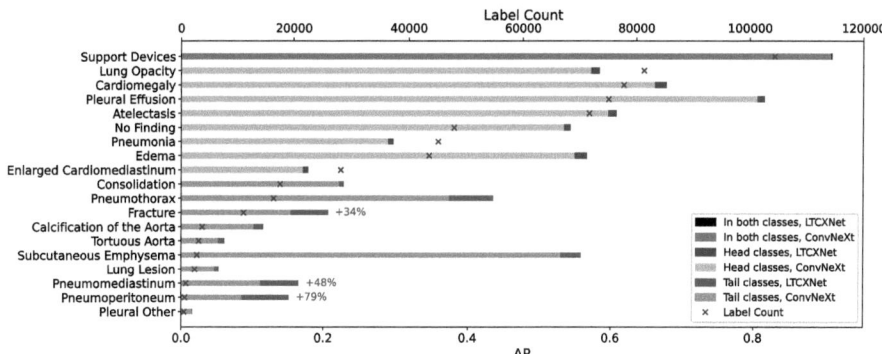

Fig. 2. Dataset label distribution and performance comparison of LTCXNet and baseline (ConvNeXt): Performance is evaluated on the testing dataset based on AP across disease conditions in CXRs, with labels sorted by frequency from high to low. The top three labels with the greatest improvement are annotated.

4 Results and Discussion

4.1 Performance Analysis on Different Disease Classes

Figure 2 presents the performance of each label. Our proposed LTCXNet achieved superior performance across all classes compared to the baseline ConvNeXt, with a particularly notable improvement in tail class performance. On average, the improvement for tail classes was 21.7%, while the performance on head classes increased by 2.3%. Among the tail classes, the three most significantly improved labels were 'Pneumoperitoneum', 'Pneumomediastinum', and 'Fracture', with performance increasing by 79%, 48%, and 34%, respectively. These results highlighted LTCXNet's effectiveness in predictions for underrepresented classes.

4.2 Task Performance Comparison with Previous Approaches

Table 1 compares previous long-tailed, multi-label classification approaches. Many degraded baseline mAP and mF1, where the baseline is LTCXNet without multi-branch learning, and compared methods use it as their backbone.

For feature decoupling in our multi-label dataset, we used a modified resampling approach to create balanced datasets, sampling until each class reached at least 70% of the rarest class's size. Five such datasets were generated for fine-tuning in the second retraining stage. However, the dataset remained imbalanced after resampling for a multi-label dataset, with the largest class being five times more frequent than the smallest one. In addition, prior research [14] showed that while Classifier Re-Training (cRT) boosted balanced accuracy on imbalanced test sets, it lowered mF1. As mAP considers precision and recall like mF1, this may explain its suboptimal performance. Learnable Weight Scaling (LWS) had similar limitations to cRT.

Table 1. Performance evaluation of different methods using resampled test sets. The mean ± std across five datasets is presented. The Baseline method is defined as LTCXNet without multi-branch learning.

Method	mAP	mF1
cRT [17]	0.348 ± 0.001	0.382 ± 0.002
LWS [17]	0.348 ± 0.001	0.383 ± 0.001
Weighted BCE Loss	0.351 ± 0.000	0.387 ± 0.001
Focal Loss [20]	0.351 ± 0.001	0.386 ± 0.002
LDAM Loss [2]	0.348 ± 0.000	0.377 ± 0.001
CB Loss [8]	0.341 ± 0.001	0.377 ± 0.001
Bsoftmax Loss [24]	0.352 ± 0.001	0.388 ± 0.002
ROS [29]	0.347 ± 0.001	0.384 ± 0.002
ML-ROS [5]	0.354 ± 0.001	0.383 ± 0.001
SimCLR [6]	0.354 ± 0.001	0.386 ± 0.002
Baseline	0.365 ± 0.001	0.384 ± 0.002
LTCXNet (Ours)	**0.378 ± 0.000**	**0.405 ± 0.001**

We compared five weighted loss variants, including weighted binary cross-entropy (BCE) loss, Focal Loss [20], Label-Distribution-Aware Margin (LDAM) Loss [2], Class-Balanced (CB) Loss [8], and Balanced Softmax (BSoftmax) Loss [24]. Configurations followed original papers, except for CB Loss, where we assigned each sample the maximum weight among its label to handle multi-label data. However, weighted loss performed suboptimally due to two reasons: over-weighting 'Tail' made the model too sensitive to rare classes, and the complex interdependencies between target labels (e.g. patients with support devices often have thoracic comorbidities) could be also disrupted by over-weighting.

ROS and ML-ROS randomly replicated samples from tail classes for a more balanced distribution. However, in multi-label datasets, this could also oversample head classes. Additionally, the overfitting from repeated samples also led to their ineffectiveness.

The performance decline in SimCLR-pretrained backbone could be attributed to two factors. First, severe class imbalances in long-tailed datasets were not adequately addressed during fine-tuning. Second, the watermarks on some CXRs biased the learned features from self-supervised learning, which lacked label information to distinguish meaningful patterns from artifacts.

4.3 Visual Explanation

Figure 3 illustrates Grad-CAM [26] visualization results of LTCXNet on three thoracic diseases, with red areas highlighting the model's focus. These visualizations align with disease characteristics. For example, Fig. 3 highlights the lower lung region, which is typically the location of pleural effusion. The locations of the heart and pneumonia patch are also accurately highlighted. These visualizations underscore the model's practical utility in identifying thoracic pathologies.

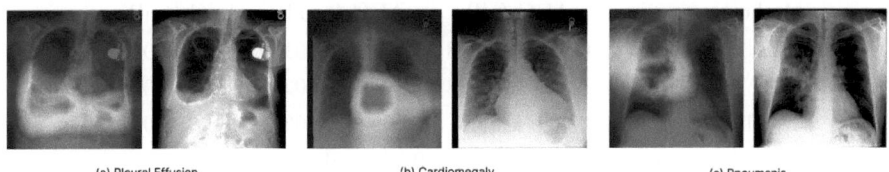

Fig. 3. Grad-CAM visualization compared to the original CXR for distinct thoracic diseases: (a) Pleural Effusion, (b) Cardiomegaly, and (c) Pneumonia. The areas where the model focused are highlighted in red. (Color figure online)

4.4 Ablation Study on mAP and mF1 Performance

Table 2 presents the outcomes of the ablation study. We observed an incremental trend in performance as each component was added. The ablation study confirmed that each component contributed positively to overall model performance.

4.5 Racial Bias Evaluation

As shown in Table 3, LTCXNet achieves the highest mAUCr scores among existing long-tail methods, with most of them underperforming even our baseline model. We hypothesize that tail classes may be more susceptible to demographic biases due to their scarcity of positive samples. To investigate this, we calculated mAUCr separately for head and tail classes (excluding Support Device from tail classes as per our multi-branch learning architecture). Results revealed largely lower mAUCr in tail classes compared to head classes. Notably, most of our model's improvement attributable to decreased racial bias in tail categories. These findings support our hypothesis that predictions for tail classes exhibit greater racial bias. Table 4 presents an ablation study analyzing individual components of LTCXNet across mAUCr(all), mAUCr(head), and mAUCr(tail). Results demonstrate that both MLDecoder and multi-branch learning effectively

Table 2. Ablation analysis of LTCXNet. The mean ± std across five datasets is presented. CNeXt: ConvNext MLD: MLDecoder. MB: Multi-branch learning.

CNeXt	MLD	MB	mAP	mF1
✓	×	×	0.350 ± 0.000	0.384 ± 0.002
✓	✓	×	0.365 ± 0.001	0.396 ± 0.001
✓	✓	✓	0.378 ± 0.000	0.405 ± 0.001

Table 3. Racial fairness evaluation of different methods using resampled test sets. The mean ± std across five datasets is presented. The Baseline method is defined as LTCXNet without multi-branch learning.

Method	mAUCr(all)	mAUCr(head)	mAUCr(tail)
cRT [17]	0.958 ± 0.008	0.971 ± 0.001	0.908 ± 0.016
LWS [17]	0.957 ± 0.008	0.971 ± 0.002	0.912 ± 0.019
Weighted BCE Loss	0.967 ± 0.001	0.969 ± 0.003	0.932 ± 0.011
Focal Loss [20]	0.972 ± 0.004	0.966 ± 0.003	0.939 ± 0.011
LDAM Loss [2]	0.966 ± 0.004	0.972 ± 0.002	0.932 ± 0.013
CB Loss [8]	0.933 ± 0.010	0.971 ± 0.001	0.857 ± 0.020
Bsoftmax Loss [24]	0.953 ± 0.006	0.972 ± 0.003	0.907 ± 0.013
ROS [29]	0.961 ± 0.006	0.971 ± 0.003	0.915 ± 0.010
ML-ROS [5]	0.953 ± 0.003	0.970 ± 0.003	0.907 ± 0.010
SimCLR [6]	0.959 ± 0.005	0.972 ± 0.003	0.922 ± 0.012
Baseline	0.967 ± 0.006	**0.973 ± 0.001**	0.927 ± 0.010
LTCXNet (Ours)	**0.973 ± 0.003**	0.971 ± 0.003	**0.946 ± 0.010**

Table 4. Ablation study on racial fairness in LTCXNet. The mean ± std across five datasets is presented. CNext: ConvNext. MLD: MLDecoder. Aug: Data augmentation. MB: Multi-branch learning.

CNext	MLD	Aug	MB	mAUCr(all)	mAUCr(head)	mAUCr(tail)
✓	✗	✗	✗	0.954 ± 0.004	0.966 ± 0.003	0.911 ± 0.009
✓	✓	✗	✗	0.960 ± 0.007	0.972 ± 0.004	0.940 ± 0.013
✓	✓	✓	✗	0.962 ± 0.007	0.973 ± 0.001	0.927 ± 0.010
✓	✓	✓	✓	0.965 ± 0.008	0.971 ± 0.001	0.946 ± 0.010

reduce mAUCr in tail classes, while image data augmentation primarily benefits head classes. While numerous methods address long-tail classification challenges, our findings emphasize the need in future study to simultaneously evaluate their differential impacts across demographic groups given the increased susceptibility of tail classes to racial bias.

5 Conclusion

LTCXNet was developed to tackle the complexities of long-tailed, multi-label classification. Our research also highlighted a potential link between racial bias and class distribution in this context. This work contributes to the advancement of trustworthy AI in medical imaging and underscores the necessity for more thorough fairness validation in existing healthcare AI methodologies.

Disclosure of Interests. The authors have no competing interests to declare that are relevant to the content of this article.

References

1. Assran, M., et al.: Self-supervised learning from images with a joint-embedding predictive architecture. In: Proceedings of the IEEE/CVF Conference on Computer Vision and Pattern Recognition, pp. 15619–15629 (2023)
2. Cao, K., Wei, C., Gaidon, A., Arechiga, N., Ma, T.: Learning imbalanced datasets with label-distribution-aware margin loss. In: Advances in Neural Information Processing Systems, vol. 32 (2019)
3. Caron, M., et al.: Emerging properties in self-supervised vision transformers. In: Proceedings of the IEEE/CVF International Conference on Computer Vision, pp. 9650–9660 (2021)
4. Castelnovo, A., Crupi, R., Greco, G., Regoli, D., Penco, I.G., Cosentini, A.C.: A clarification of the nuances in the fairness metrics landscape. Sci. Rep. **12**(1), 4209 (2022)
5. Charte, F., Rivera, A.J., Jesus, M.J., Herrera, F.: Addressing imbalance in multilabel classification: measures and random resampling algorithms. Neurocomputing **163**, 3–16 (2015)
6. Chen, T., Kornblith, S., Norouzi, M., Hinton, G.: A simple framework for contrastive learning of visual representations. In: International Conference on Machine Learning, pp. 1597–1607. PMLR (2020)
7. Chen, X., et al.: Symbolic discovery of optimization algorithms (2023). https://arxiv.org/abs/2302.06675
8. Cui, Y., Jia, M., Lin, T.Y., Song, Y., Belongie, S.: Class-balanced loss based on effective number of samples. In: Proceedings of the IEEE/CVF Conference on Computer Vision and Pattern Recognition, pp. 9268–9277 (2019)
9. Deng, J., Dong, W., Socher, R., Li, L.J., Li, K., Fei-Fei, L.: ImageNet: a large-scale hierarchical image database. In: 2009 IEEE Conference on Computer Vision and Pattern Recognition, pp. 248–255 (2009). https://doi.org/10.1109/CVPR.2009.5206848
10. Everingham, M., Gool, L., Williams, C.K., Winn, J., Zisserman, A.: The pascal visual object classes (VOC) challenge. Int. J. Comput. Vision **88**, 303–338 (2010)
11. Gichoya, J.W., et al.: AI recognition of patient race in medical imaging: a modelling study. Lancet Digit. Health **4**(6), e406–e414 (2022)
12. Holste, G., et al.: How does pruning impact long-tailed multi-label medical image classifiers? In: Greenspan, H., et al. (eds.) MICCAI 2023. LNCS, vol. 14224, pp. 663–673. Springer, Cham (2023). https://doi.org/10.1007/978-3-031-43904-9_64
13. Holste, G., et al.: CXR-LT: multi-label long-tailed classification on chest X-rays. PhysioNet **5**, 19 (2023)
14. Holste, G., et al.: Long-tailed classification of thorax diseases on chest X-ray: a new benchmark study. In: Nguyen, H.V., Huang, S.X., Xue, Y. (eds.) DALI 2022. LNCS, vol. 13567, pp. 22–32. Springer, Cham (2022). https://doi.org/10.1007/978-3-031-17027-0_3
15. Jeong, J., Jeoun, B., Park, Y., Han, B.: An optimized ensemble framework for multi-label classification on long-tailed chest X-ray data. In: Proceedings of the IEEE/CVF International Conference on Computer Vision, pp. 2739–2746 (2023)

16. Kälsch, H., et al.: Aortic calcification onset and progression: association with the development of coronary atherosclerosis. J. Am. Heart Assoc. **6**(4), e005093 (2017)
17. Kang, B., et al.: Decoupling representation and classifier for long-tailed recognition. arXiv preprint arXiv:1910.09217 (2019)
18. Lanchantin, J., Wang, T., Ordonez, V., Qi, Y.: General multi-label image classification with transformers. In: Proceedings of the IEEE/CVF Conference on Computer Vision and Pattern Recognition, pp. 16478–16488 (2021)
19. Li, B., Han, Z., Li, H., Fu, H., Zhang, C.: Trustworthy long-tailed classification. In: Proceedings of the IEEE/CVF Conference on Computer Vision and Pattern Recognition, pp. 6970–6979 (2022)
20. Lin, T.Y., Goyal, P., Girshick, R., He, K., Dollár, P.: Focal loss for dense object detection. In: Proceedings of the IEEE International Conference on Computer Vision, pp. 2980–2988 (2017)
21. Liu, Z., Mao, H., Wu, C.Y., Feichtenhofer, C., Darrell, T., Xie, S.: A convnet for the 2020s. In: Proceedings of the IEEE/CVF Conference on Computer Vision and Pattern Recognition, pp. 11976–11986 (2022)
22. Paszke, A., et al.: PyTorch: an imperative style, high-performance deep learning library. In: Advances in Neural Information Processing Systems, vol. 32 (2019)
23. Rath, A., Mishra, D., Panda, G., Satapathy, S.C.: Heart disease detection using deep learning methods from imbalanced ECG samples. Biomed. Signal Process. Control **68**, 102820 (2021)
24. Ren, J., Yu, C., Ma, X., Zhao, H., Yi, S., et al.: Balanced meta-softmax for long-tailed visual recognition. In: Advances in Neural Information Processing Systems, vol. 33, pp. 4175–4186 (2020)
25. Ridnik, T., Sharir, G., Ben-Cohen, A., Ben-Baruch, E., Noy, A.: ML-Decoder: scalable and versatile classification head. In: Proceedings of the IEEE/CVF Winter Conference on Applications of Computer Vision, pp. 32–41 (2023)
26. Selvaraju, R.R., Cogswell, M., Das, A., Vedantam, R., Parikh, D., Batra, D.: Grad-CAM: visual explanations from deep networks via gradient-based localization. In: Proceedings of the IEEE International Conference on Computer Vision, pp. 618–626 (2017)
27. Song, Y., Zhou, Q., Hu, K., Ma, L., Lu, X.: CFRL: coarse-fine decoupled representation learning for long-tailed recognition. In: Proceedings of the 6th ACM International Conference on Multimedia in Asia, pp. 1–7 (2024)
28. Subramanian, S., Rahimi, A., Baldwin, T., Cohn, T., Frermann, L.: Fairness-aware class imbalanced learning. arXiv preprint arXiv:2109.10444 (2021)
29. Tarekegn, A.N., Giacobini, M., Michalak, K.: A review of methods for imbalanced multi-label classification. Pattern Recogn. **118**, 107965 (2021)
30. Wang, Z., Liu, J.C.: Translating math formula images to latex sequences using deep neural networks with sequence-level training (2019)
31. Yan, S., Kao, H.T., Ferrara, E.: Fair class balancing: enhancing model fairness without observing sensitive attributes. In: Proceedings of the 29th ACM International Conference on Information & Knowledge Management, pp. 1715–1724 (2020)

Fairness and Robustness of CLIP-Based Models for Chest X-Rays

Théo Sourget[1(✉)], David Restrepo[1], Céline Hudelot[1], Enzo Ferrante[2], Stergios Christodoulidis[1], and Maria Vakalopoulou[1]

[1] MICS, CentraleSupélec - Université Paris-Saclay, Gif-sur-Yvette, France
{theo.sourget,david.restrepo,celine.hudelot,
stergios.christodoulidis,maria.vakalopoulou}@centralesupelec.fr
[2] CONICET, Universidad de Buenos Aires, Buenos Aires, Argentina
eferrante@sinc.unl.edu.ar

Abstract. Motivated by the strong performance of CLIP-based models in natural image-text domains, recent efforts have adapted these architectures to medical tasks, particularly in radiology, where large paired datasets of images and reports, such as chest X-rays, are available. While these models have shown encouraging results in terms of accuracy and discriminative performance, their fairness and robustness in the different clinical tasks remain largely underexplored. In this study, we extensively evaluate six widely used CLIP-based models on chest X-ray classification using three publicly available datasets: MIMIC-CXR, NIH-CXR14, and NEATX. We assess the models fairness across six conditions and patient subgroups based on age, sex, and race. Additionally, we assess the robustness to shortcut learning by evaluating performance on pneumothorax cases with and without chest drains. Our results indicate performance gaps between patients of different ages, but more equitable results for the other attributes. Moreover, all models exhibit lower performance on images without chest drains, suggesting reliance on spurious correlations. We further complement the performance analysis with a study of the embeddings generated by the models. While the sensitive attributes could be classified from the embeddings, we do not see such patterns using PCA, showing the limitations of these visualisation techniques when assessing models. Our code is available at https://github.com/TheoSourget/clip_cxr_fairness.

Keywords: CLIP-based models · Chest X-ray · Fairness · Shortcut

1 Introduction

Deep learning models that have been trained on large-scale chest X-ray datasets have achieved performances reaching expert-levels in disease classification of X-ray images [10,22]. However, despite such models obtaining strong benchmark performance, different studies show that they often exhibit performance disparities across patient subgroups, revealing concerning biases. For example, studies like [9,14] show how the performance of convolutional neural networks (CNN) can vary based on demographic attributes such as age, sex, or race, particularly for X-rays. Beyond performance differences, there is growing evidence that deep learning models encode sensitive demographic information in their internal representations. Previous studies like [1,19] are able to predict sensitive attributes from the embedding generated by pretrained models. Similarly, Gichoya et al. [8] demonstrate that deep learning models can classify patient race, even when the input images are heavily corrupted, raising serious concerns about the implicit encoding of sensitive information and its fairness implications.

Additionally, other studies show the impact of artefacts, also called shortcuts, in the classification. Jiménez-Sánchez et al. [11] and Oakden et al. [18] show that models for pneumothorax classification have lower performances on images without chest drains, a common treatment for this disease. Moreover, Sourget et al. [23] demonstrate the ability of the models to obtain good performances in chest X-rays classification while masking out the lungs in the image, showing how these models can rely on non-relevant features.

More recently, advances in multimodal and foundation models have led to the development of contrastively trained architectures that jointly leverage chest X-rays and radiology reports [2,3,5,24,26,28]. While these vision-language models (VLMs) have demonstrated promising results, recent studies have raised concerns regarding the fairness of VLMs. Luo et al. [16] assess the fairness of the original CLIP and BLIP2 models on glaucoma classification pretrained with both natural domain and medical data, showing differences across subgroups especially on the natural domain models. Yang et al. [27] evaluate the fairness of the CheXzero model [24] for chest X-ray classification, showing the gap in performances between different subgroups. Finally, Fay et al. [7] compare the performances of multiple zero-shot and training-based strategies for the MedImageInsight model [5] on pneumonia classification and include an assessment of their fairness, showing that zero-shot techniques present less bias compared to linear probing but still higher than with LoRA or k-NN.

In this work, we extend these studies by evaluating a large set of CLIP-based models pretrained on X-ray data, providing complementary analysis of the embedding representations, and assessing the robustness of the models to shortcut learning. Our empirical study, aiming at improving the understanding of the biases of CLIP-based models for chest X-rays classification: **1)** evaluates the performance of six widely used CLIP-based vision-language models on the multilabel classification of chest X-rays; **2)** assesses the fairness of the architectures on multiple subgroups of patients; **3)** studies the potential encoding of sensitive attributes in the embedding of these models with visualisation and classification techniques; **4)** compares their robustness regarding shortcuts on pneumothorax classification with and without chest drains.

2 Fairness and Robustness of CLIP-Based Models

2.1 Data

In this study, we use three public datasets for X-Rays. The MIMIC-CXR[1] dataset contains chest X-rays and radiology reports from 227,835 radiographic studies. Following standard practices in the training and evaluation of foundation models, we only use the original test split containing 30,359 images to avoid potential data leakage. Since some of our analyses need the "FINDING" section of the report, we only kept the 8950 samples for which this section is available in the test set. We use a subset of the classes available in the dataset: atelectasis, cardiomegaly, consolidation, pleural effusion, pneumonia, and pneumothorax.

The NIH-CXR14 dataset[2] contains 112,120 X-ray images from 30,805 unique patients. We only use the 25,596 images of the test set. While the dataset contains annotations of 14 different conditions, here we focus on pneumothorax for our shortcut learning analysis. The NEATX dataset[3] contains annotations of chest drains in X-rays from the NIH-CXR14 and PadChest datasets. We use the annotations for the NIH-CXR14 dataset to assess the robustness of models to chest drains in pneumothorax classification. As the NEATX dataset only contains annotations of chest drains in positive samples of pneumothorax, we train a DenseNet model for the detection of chest drains using the hyperparameters described in the dataset paper [6] and automatically generate the labels for non-pneumothorax samples.

2.2 Models

We conduct our experiments with six CLIP-based architectures for which pre-trained weights were available: MedCLIP [26], Biovil [3] and Biovil-t [2], MedImageInsight [5], CheXzero [24], and CXR-CLIP [28]. All of these models were trained on datasets containing chest X-rays either exclusively or with other medical image modalities. We selected these models due to their recent release and their wide usage as baseline in previous works. CLIP-based architectures consist of two parts: an image encoder and a text encoder, trained using contrastive learning to align the embeddings of image and text pairs.

2.3 Evaluation Protocol on Zero-Shot Classification

We evaluate the performance of the models on different subgroups in a zero-shot classification setting. Inspired by the setups of [7,24], we compute for each label the cosine similarities between the embeddings of an image and two templates "Chest {CLASS}" and "Chest No Findings". We then apply a softmax

---

[1] Downloaded from https://physionet.org/content/mimic-cxr-jpg/2.1.0/ and complemented with MIMIC-IV: https://physionet.org/content/mimiciv/3.1/.
[2] Version 3 downloaded from https://www.kaggle.com/datasets/nih-chest-xrays/data.
[3] Version 1.0 downloaded from https://zenodo.org/records/14944064.

function between the two similarities to obtain the probability of the disease. We evaluate the models across individual diseases and demographic subgroups (sex, race, and age) to assess both overall discriminative performance and subgroup fairness, quantified by performance disparities. We also assess whether the models rely on non-clinically relevant image features. To this end, we evaluate their performance for pneumothorax classification on two groups: one in which all patients with pneumothorax have chest drains, and the other in which they never have one. Finally, we compute calibration curves for this task using the softmax values from the zero-shot classification to examine the reliability of the predicted probabilities.

As metrics, we use the area under the receiver operating characteristic (AUC) and the adjusted area under the precision-recall curve ($AUPRC_{adj}$), which is usually adopted to evaluate models in a highly imbalanced scenario [17]. The $AUPRC_{adj}$ is defined as $1 - \frac{log(AUPRC)}{log(AUPRC_{rng})}$ with $AUPRC_{rng}$ being the ratio between the number of positive samples for a class and the total number of samples. We also compute 95% confidence intervals using the bootstrap method with 1000 resamples. During a resample, we draw n samples with replacement from the base evaluation set to compute the metrics, n being the number of samples from the base group.

2.4 Encoding of Sensitive Attributes in the Embedding Space

To further understand how these models work and what they learn in this multimodal contrastive setting, we generate and visualise the obtained image and text embeddings. For textual embeddings, as the text encoders have a limited input size, we only use the "FINDINGS" section from radiology reports, likely to contain the most relevant information. To assess the encoding of sensitive attributes, we use PCA to project the embeddings in two dimensions, revealing potential patterns with respect to patient sex, race, and age. We also train a model to classify the different sensitive attributes from the image embeddings using simple models like a linear probe (LP), a k-nearest neighbours (k-NN) classifier and a single-hidden-layer multi-layer perceptron (MLP). We split the original test sets in train, validation, and test subsets, ensuring that all images from a given patient are assigned to the same split to prevent data leakage. We used the validation set to tune the models' hyperparameters: the learning rate in the linear probe, the number of nearest neighbours, and the number of neurons of the MLP hidden layer.

Finally, following the analysis by Schrodi et al. [20] on the modality gap—which shows that differences between image and text embedding centroids are concentrated in a few dimensions—we conduct a similar analysis across patient subgroups. Specifically, we compute the centroid of image embeddings for each subgroup and measure the per-dimension differences between pairs of subgroup centroids (e.g. the centroids of female and male patients). These differences are then ordered in descending order to assess the potential gap between the embeddings of two subgroups and how many dimensions contribute to it.

3 Results

3.1 Good Overall Performances with Subgroup-Specific Variability

Table 1 shows the AUC and AUPRC$_{adj}$ of the different models on the MIMIC-CXR test set. One can see that aside from CXR-CLIP, the models obtain better than random values, especially for MedCLIP, MedImageInsight, and CheXzero, confirming their application in zero-shot settings.

For further evaluation, we generate for each model a barplot of the AUC and AUPRC$_{adj}$ per subgroup to observe potential gaps, see the results for the MedCLIP model in Fig. 1 with 95% confidence intervals computed using the bootstrap method. While the results vary across the models and subgroups, we can still see a similar pattern with gaps across patient ages. The gaps seem, however, smaller for patient sex and race, with the exception of Asian patients, for which we can often see either a high improvement or decrease. However, this may be explained by the limited amount of positive samples per class for Asian patients, leading to more extreme values and confidence intervals. Note that the same observation can be made for the 18–25 year old subgroup. This observation highlights the need for a more diverse test dataset to better estimate the true performance of the models on these subgroups.

Table 1. AUC and AUPRC$_{adj}$ of zeroshot classification. Negative AUPRC$_{adj}$ values denote results below the random classifier. Values in [] are the 95% confidence intervals computed with the bootstrap method. ± in the Mean column are the standard deviations.

	Atelectasis		Cardiomegaly		Consolidation		Effusion		Pneumonia		Pneumothorax		Mean	
	AUC	AUPRC$_{adj}$	AUC	AUPRC$_{adj}$	AUC	AUPRC$_{adj}$	AUC	AUPRC$_{adj}$	AUC	AUPRC$_{adj}$	AUC	AUPRC$_{adj}$	AUC	AUPRC$_{adj}$
MedCLIP	**0.8**	**0.54**	**0.8**	0.52	**0.84**	0.4	**0.92**	**0.81**	**0.74**	0.46	**0.88**	**0.74**	**0.83**	**0.58**
	[0.79, 0.82]	[0.51, 0.57]	[0.78, 0.81]	[0.48, 0.56]	[0.82, 0.86]	[0.36, 0.46]	[0.91, 0.93]	[0.79, 0.83]	[0.72, 0.76]	[0.41, 0.51]	[0.86, 0.91]	[0.7, 0.78]	± 0.06	± 0.16
Biovil	0.68	0.2	0.76	0.41	0.42	−0.06	0.69	0.38	0.49	−0.0	0.72	0.21	0.63	0.19
	[0.66, 0.69]	[0.18, 0.23]	[0.74, 0.77]	[0.36, 0.45]	[0.39, 0.46]	[−0.08, −0.03]	[0.67, 0.7]	[0.35, 0.42]	[0.47, 0.52]	[−0.03, 0.03]	[0.69, 0.74]	[0.18, 0.25]	± 0.14	± 0.19
Biovil-t	0.64	0.15	0.74	0.3	0.59	0.06	0.79	0.49	0.61	0.14	0.66	0.17	0.67	0.22
	[0.63, 0.66]	[0.13, 0.18]	[0.73, 0.76]	[0.27, 0.34]	[0.55, 0.62]	[0.03, 0.11]	[0.78, 0.8]	[0.46, 0.52]	[0.58, 0.63]	[0.11, 0.19]	[0.63, 0.69]	[0.13, 0.21]	± 0.08	± 0.15
MedImageInsight	0.74	0.36	**0.85**	**0.53**	0.83	0.4	0.88	0.7	0.69	0.33	**0.88**	0.63	0.81	0.49
	[0.73, 0.75]	[0.33, 0.39]	[0.83, 0.86]	[0.49, 0.57]	[0.8, 0.85]	[0.35, 0.46]	[0.87, 0.89]	[0.67, 0.72]	[0.67, 0.72]	[0.29, 0.38]	[0.86, 0.9]	[0.58, 0.68]	± 0.08	± 0.15
CheXzero	0.67	0.21	**0.85**	0.6	0.8	0.34	0.88	0.7	0.68	0.28	0.8	0.36	0.78	0.42
	[0.65, 0.68]	[0.19, 0.25]	[0.83, 0.86]	[0.57, 0.64]	[0.78, 0.83]	[0.29, 0.4]	[0.87, 0.89]	[0.68, 0.72]	[0.66, 0.7]	[0.24, 0.33]	[0.78, 0.82]	[0.31, 0.43]	± 0.09	± 0.19
CXR-CLIP	0.61	0.18	0.55	0.06	0.48	−0.01	0.67	0.35	0.48	−0.02	0.45	−0.01	0.54	0.09
	[0.59, 0.63]	[0.15, 0.22]	[0.53, 0.58]	[0.03, 0.09]	[0.46, 0.49]	[0.0, −0.0]	[0.66, 0.69]	[0.31, 0.38]	[0.46, 0.5]	[0.0, 0.02]	[0.41, 0.48]	[0.0, 0.09]	± 0.09	± 0.15

3.2 Sensitive Attributes are Encoded in Embeddings Despite Unclear Visual Separation

Even though CLIP-based architectures align image and text embeddings using contrastive learning, a simple PCA analysis reveals that in most models (MedImageInsight, CheXzero, CXR-CLIP, and MedCLIP) there is a pronounced gap between the embeddings generated by the image and text encoders. Visible in Fig. 2a, this is aligned with the results from previous studies in natural images [15,20]. Moreover, as shown in [20], we also found in Fig. 3a that the gap between the modalities is concentrated on few dimensions.

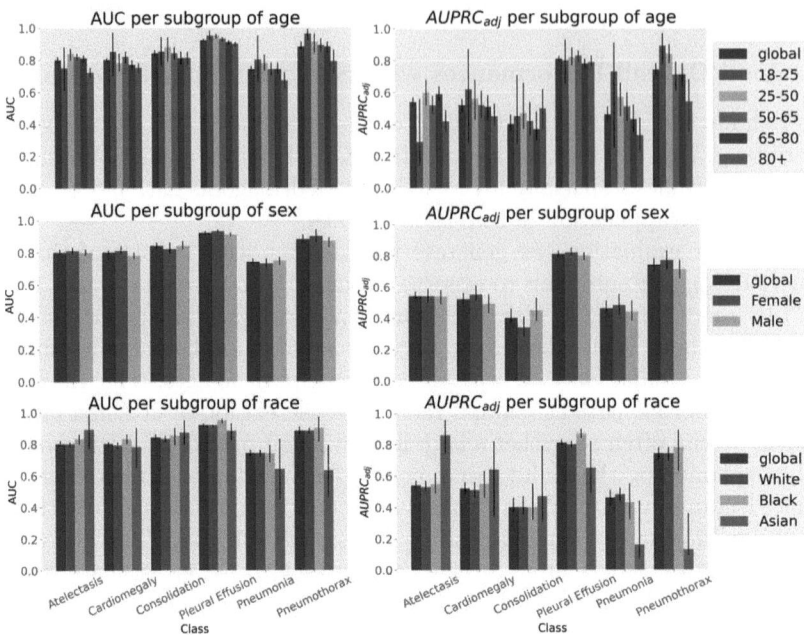

Fig. 1. Example of AUC and AUPRC$_{adj}$ for the MedCLIP model and different subgroups with 95% confidence interval using the bootstrap method.

On the other hand, as shown in Fig. 2b–2d, we do not see clear patterns in the PCA plots coloured by sensitive attributes. Instead, we observe that the different attributes seem to be well spread across the feature space in both image and text spaces. We may conclude from these visualisations that the information is not present in the embedding. However, the differences in subgroup performance observed in the previous section (particularly for age) suggest that certain information related to sensitive attributes may, in fact, be encoded in these representations. To further confirm the algorithmic encoding of protected attributes, we tested the ability of simple supervised models like linear probing, k-nearest neighbours, and MLP to classify sensitive attributes from the embeddings of each CLIP-based model and present the results in Table 2. While the MLP obtains higher performances, we observe that on patient sex and age all models are able to obtain results above random. However, we can see that for the patient race, k-NN classifiers obtained near-random results for almost all the models. The linear probe is also unable to classify the attribute for some models while the MLP still performs correctly on this attribute. This shows that while it is probably less distinguishable than the other two attributes, it may still be present in the embedding. It is important to note that while such results may show the encoding of information in the embeddings, it is not enough to conclude that they are actually used as shortcuts for other downstream tasks.

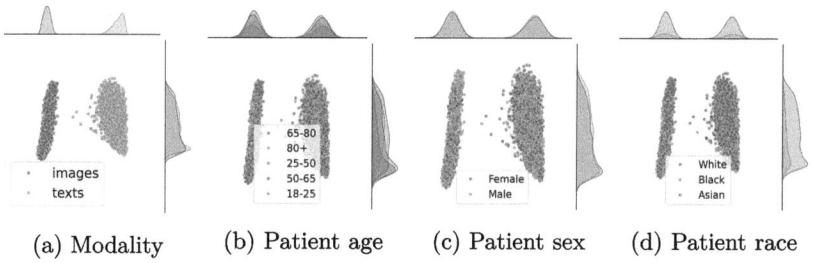

(a) Modality (b) Patient age (c) Patient sex (d) Patient race

Fig. 2. PCA of MedImageInsight image and text embeddings grouped on different attributes.

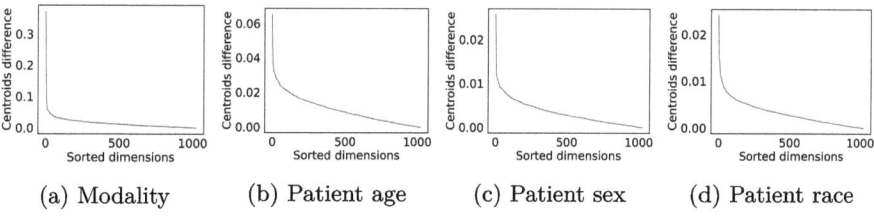

(a) Modality (b) Patient age (c) Patient sex (d) Patient race

Fig. 3. Ordered differences between each dimension of the centroids of (a) image and text embeddings generated with MedImageInsight, (b) image embeddings of 18–25 years old and 80+ years old patients, (c) image embeddings of female and male patients, and (d) image embeddings of white and black patients. Note that the y-axis range is different in the figures.

As for the modality gap, we analyse the difference between each dimension of the image embeddings centroids between two subgroups (defined by different sensitive attributes) using only the image modality. Examples are presented in Fig. 3b–3d. We found that in this case, the differences are much smaller than for the modality gap and more spread across the dimensions. These results suggest that while mitigating the modality gap can be done by focusing on few dimensions, the mitigation of subgroups biases may require more global techniques.

3.3 Evidence of Shortcut Learning and Miscalibration in CLIP-Based Models

Figure 4a and 4b show the results on chest X-ray with and without chest drains. We can see that all models except CXR-CLIP obtain better adjusted AUPRC on images with chest drains compared to X-rays without drains (ranging from +0.09 to +0.30), aligned with previous results on CNN models [11,18]. Moreover, we present the calibration curves of the models in Fig. 4c. We see that while MedImageInsight is the most calibrated model, all the models seem to be miscalibrated and overconfident. Interestingly, we can see that for CXR-CLIP, CheXzero and MedCLIP, all the probabilities are around 0.5. Despite this, both MedCLIP and CheXzero achieve acceptable AUC scores, indicating that their

Table 2. Mean AUC of sensitive attributes classification from image embeddings with a linear probe, k-NN and MLP

	Sex			Race			Age		
	LP	k-NN	MLP	LP	k-NN	MLP	LP	k-NN	MLP
MedCLIP	0.59	0.71	0.94	0.67	0.55	0.75	0.75	0.65	0.80
Biovil	0.79	0.62	0.88	0.45	0.54	0.70	0.76	0.54	0.77
Biovil-t	0.65	0.56	0.86	0.54	0.54	0.67	0.65	0.62	0.78
MedImageInsight	0.98	0.82	0.97	0.78	0.62	0.80	0.87	0.77	0.84
CheXzero	0.75	0.65	0.93	0.66	0.53	0.69	0.76	0.69	0.78
CXR-CLIP	0.97	0.89	0.97	0.82	0.55	0.80	0.84	0.58	0.80

discriminative performance remains unaffected, likely because samples are still correctly ranked within this narrow probability range. However, this behaviour significantly complicates the interpretability of individual predictions.

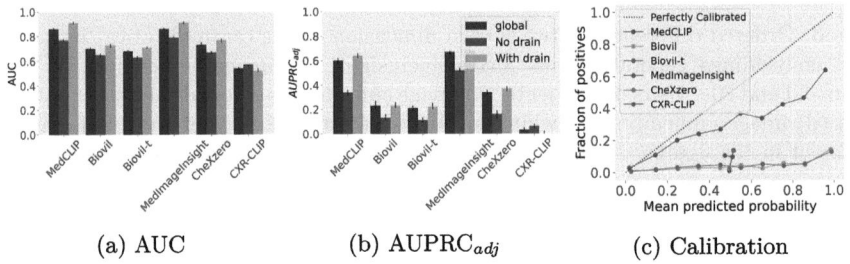

(a) AUC (b) AUPRC$_{adj}$ (c) Calibration

Fig. 4. a) AUC and (b) adjusted AUPRC of all models on pneumothorax classification of chest X-rays with and without chest drains. (c) Calibration curves of the models on all images.

4 Discussion and Conclusions

In this study, we analysed the fairness of CLIP-based models for chest X-rays across multiple subgroups of patients, showing gaps in the performance obtained on patients of various ages but more balanced results on the other attributes. We evaluated the robustness of the models to shortcut learning using chest drains in pneumothorax classification and showed that all models had a better AUPRC$_{adj}$ on images with drains compared to images without drains, indicating their potential reliance on this spurious correlation. We also assessed the calibrations of the models and found that all models were miscalibrated. In addition to the performances, we studied the embeddings generated by the models and found that while we could not discover encoding of sensitive attributes using

PCA visualizations, such attributes could still be classified from the embeddings using simple supervised models like k-NN or MLP, suggesting the encoding of protected attributes.

Fairness and robustness analyses are often constrained by the specific datasets and models used in a given study. Although our experiments span multiple datasets and CLIP-based architectures, the findings may not generalize to all CLIP variants or to other datasets. This highlights the need for similar evaluations in diverse contexts. Furthermore, our fairness analysis focuses on single sensitive attributes at a time, whereas prior research has shown that intersecting subgroups (e.g., age and race) can expose additional fairness concerns [21]. Lastly, since not all multimodal models follow the CLIP framework, extending such evaluations to generative instead of contrastive multimodal models would provide a more comprehensive understanding.

Our findings, supported by prior work, underscore the importance of improved evaluation frameworks to assess not only overall performance but also fairness across patient subgroups. This requires more diverse datasets, the use of task-appropriate metrics, and going beyond accuracy to consider aspects like calibration and bias.

Acknowledgements. This work was partially funded by the RHU-EndoVx project (21-RHUS-0011, ANR), Hagnodice project (ANR-21-CE45-0007). This work was performed using HPC resources from the Mesocentre computing center of CentraleSupelec. EF was supported by the Google Award for Inclusion Research and a Googler Initiated Grant. EF and MV are supported by the STIC-AmSud CGFLRVE project. We want to thank the providers of the MIMIC-CXR, NIH-CXR14 and NEATX for creating the datasets used in this study.

References

1. Bahre, G.H., Hamidi, H., Calimeri, F., Sellergren, A., Celi, L.A., Seyyed-Kalantari, L.: Fairness of AI models in vector embedded chest X-ray representations. In: Advancements in Medical Foundation Models: Explainability, Robustness, Security, and Beyond (2024)
2. Bannur, S., et al.: Learning to exploit temporal structure for biomedical vision-language processing. In: Proceedings of the IEEE/CVF Conference on Computer Vision and Pattern Recognition, pp. 15016–15027 (2023)
3. Boecking, B., et al.: Making the most of text semantics to improve biomedical vision–language processing. In: Avidan, S., Brostow, G., Cissé, M., Farinella, G.M., Hassner, T. (eds.) ECCV 2022. LNCS, vol. 13696, pp. 1–21. Springer, Cham (2022). https://doi.org/10.1007/978-3-031-20059-5_1
4. Cheplygina, V., Cathrine, D., Eriksen, T.N., Jiménez-Sánchez, A.: NEATX: non-expert annotations of tubes in X-rays (2025). https://doi.org/10.5281/zenodo.14944064
5. Codella, N.C., et al.: MedImageInsight: an open-source embedding model for general domain medical imaging. arXiv preprint arXiv:2410.06542 (2024)

6. Damgaard, C., Eriksen, T.N., Juodelyte, D., Cheplygina, V., Jiménez-Sánchez, A.: Augmenting chest X-ray datasets with non-expert annotations. arXiv preprint arXiv:2309.02244 (2023)
7. Fay, L., et al.: Beyond the prompt: deploying medical foundation models on diverse chest X-ray populations. In: Medical Imaging with Deep Learning (2025). https://openreview.net/forum?id=RuqEg2XAWq
8. Gichoya, J.W., et al.: AI recognition of patient race in medical imaging: a modelling study. Lancet Digit. Health **4**(6), e406–e414 (2022)
9. Glocker, B., Jones, C., Bernhardt, M., Winzeck, S.: Algorithmic encoding of protected characteristics in chest X-ray disease detection models. EBioMedicine **89** (2023)
10. Hosny, A., Parmar, C., Quackenbush, J., Schwartz, L.H., Aerts, H.J.: Artificial intelligence in radiology. Nat. Rev. Cancer **18**(8), 500–510 (2018)
11. Jiménez-Sánchez, A., Juodelyte, D., Chamberlain, B., Cheplygina, V.: Detecting shortcuts in medical images-a case study in chest X-rays. In: 2023 IEEE 20th International Symposium on Biomedical Imaging (ISBI), pp. 1–5. IEEE (2023)
12. Johnson, A.E., et al.: MIMIC-CXR, a de-identified publicly available database of chest radiographs with free-text reports. Sci. Data **6**(1), 317 (2019)
13. Johnson, A.E., Pollard, T.J., Mark, R.G., Berkowitz, S.J., Horng, S.: MIMIC-CXR database (version 2.1.0). PhysioNet (2024). https://doi.org/10.13026/4jqj-jw95
14. Larrazabal, A.J., Nieto, N., Peterson, V., Milone, D.H., Ferrante, E.: Gender imbalance in medical imaging datasets produces biased classifiers for computer-aided diagnosis. Proc. Natl. Acad. Sci. **117**(23), 12592–12594 (2020)
15. Liang, V.W., Zhang, Y., Kwon, Y., Yeung, S., Zou, J.Y.: Mind the gap: understanding the modality gap in multi-modal contrastive representation learning. In: Advances in Neural Information Processing Systems, vol. 35, pp. 17612–17625 (2022)
16. Luo, Y., et al.: FairCLIP: harnessing fairness in vision-language learning. In: Proceedings of the IEEE/CVF Conference on Computer Vision and Pattern Recognition, pp. 12289–12301 (2024)
17. Mosquera, C., Ferrer, L., Milone, D.H., Luna, D., Ferrante, E.: Class imbalance on medical image classification: towards better evaluation practices for discrimination and calibration performance. Eur. Radiol. **34**(12), 7895–7903 (2024)
18. Oakden-Rayner, L., Dunnmon, J., Carneiro, G., Ré, C.: Hidden stratification causes clinically meaningful failures in machine learning for medical imaging. In: Proceedings of the ACM Conference on Health, Inference, and Learning, pp. 151–159 (2020)
19. Restrepo, D., Wu, C., Vásquez-Venegas, C., Nakayama, L.F., Celi, L.A., López, D.M.: DF-DM: a foundational process model for multimodal data fusion in the artificial intelligence era. Res. Square, rs–3 (2024)
20. Schrodi, S., Hoffmann, D.T., Argus, M., Fischer, V., Brox, T.: Two effects, one trigger: on the modality gap, object bias, and information imbalance in contrastive vision-language models. In: The Thirteenth International Conference on Learning Representations (2025). https://openreview.net/forum?id=uAFHCZRmXk
21. Seyyed-Kalantari, L., Zhang, H., McDermott, M.B., Chen, I.Y., Ghassemi, M.: Underdiagnosis bias of artificial intelligence algorithms applied to chest radiographs in under-served patient populations. Nat. Med. **27**(12), 2176–2182 (2021)
22. Shen, J., et al.: Artificial intelligence versus clinicians in disease diagnosis: systematic review. JMIR Med. Inform. **7**(3), e10010 (2019). https://doi.org/10.2196/10010

23. Sourget, T., Hestbek-Møller, M., Jiménez-Sánchez, A., Junchi Xu, J., Cheplygina, V.: Mask of truth: model sensitivity to unexpected regions of medical images. J. Imaging Inf. Med., 1–18 (2025)
24. Tiu, E., Talius, E., Patel, P., Langlotz, C.P., Ng, A.Y., Rajpurkar, P.: Expert-level detection of pathologies from unannotated chest X-ray images via self-supervised learning. Nat. Biomed. Eng. **6**(12), 1399–1406 (2022)
25. Wang, X., Peng, Y., Lu, L., Lu, Z., Bagheri, M., Summers, R.M.: ChestX-ray8: hospital-scale chest X-ray database and benchmarks on weakly-supervised classification and localization of common thorax diseases. In: Computer Vision and Pattern Recognition, pp. 2097–2106 (2017)
26. Wang, Z., Wu, Z., Agarwal, D., Sun, J.: MedCLIP: contrastive learning from unpaired medical images and text. In: Proceedings of the Conference on Empirical Methods in Natural Language Processing. Conference on Empirical Methods in Natural Language Processing, vol. 2022, p. 3876 (2022)
27. Yang, Y., et al.: Demographic bias of expert-level vision-language foundation models in medical imaging. Sci. Adv. **11**(13), eadq0305 (2025)
28. You, K., et al.: CXR-CLIP: toward large scale chest X-ray language-image pretraining. In: Greenspan, H., et al. (eds.) MICCAI 2023. LNCS, vol. 14221, pp. 101–111. Springer, Cham (2023). https://doi.org/10.1007/978-3-031-43895-0_10

ShortCXR: Benchmarking Self-supervised Learning for Shortcut Mitigation in Chest X-Ray Diagnostics

You-Qi Chang-Liao and Po-Chih Kuo(✉)

National Tsing Hua University, Hsinchu, Taiwan
kuopc@cs.nthu.edu.tw

Abstract. Ensuring both accuracy and fairness in AI-assisted chest X-ray (CXR) diagnostics is critical, particularly given the presence of sensitive patient information. In this study, we introduce ShortCXR, a Shortcut-Infused Chest X-Ray Dataset explicitly designed to expose and measure biases stemming from intrinsic image features, data sources, and demographic factors. We then integrate self-supervised learning (SSL) techniques into a series of experiments to evaluate how these biases affect diagnostic performance. Our findings demonstrate that state-of-the-art SSL methods boost both accuracy and fairness, effectively mitigating the impact of distortion-based, source-based, and demographic biases. By highlighting SSL's potential to enhance diagnostic quality and equity, this work provides a strong foundation for future research on bias mitigation in medical imaging.

Keywords: Fairness · Shortcut · Self-supervised learning · Chest X-ray

1 Introduction

Deep neural networks have significantly advanced across numerous applications; however, their reliability and transparency remain pressing concerns. In particular, networks may learn *shortcuts*—spurious correlations or hidden cues in training data—that artificially inflate performance yet fail to generalize to new settings. These shortcuts often go undetected because they exploit subtle dataset artifacts or background elements, ultimately undermining trust in AI systems.

Such reliability issues are particularly consequential in medical imaging, where patient outcomes hinge on accurate diagnoses. Chest radiography (CXR) interpretation provides a critical example: previous studies [9] have identified shortcuts—such as hospital tokens or shoulder positioning—that inadvertently influence diagnostic results. Likewise, the synthetic generation of COVID-19 positive images from negative ones using CycleGANs [24] has revealed consistent image alterations, including changes to laterality markers or peripheral radiopacities, that obscure genuine pathological cues. Furthermore, a large-scale investigation [4] emphasized the complexities of ensuring equitable diagnostics by revealing bias in CXR datasets. Other works [10,22] highlight how demographic features can serve as shortcuts, potentially compromising fairness when models exhibit reduced accuracy or higher false-positive rates for specific population groups.

Efforts to counter these biases and enhance robustness against out-of-distribution (o.o.d.) data have led to a variety of mitigation strategies. Among these, self-supervised learning (SSL) [14] has gained traction by learning meaningful representations from unlabeled data using pretext tasks. Early SSL approaches such as Jigsaw [19] or Rotations [11] have evolved into more powerful contrastive methods, including BYOL [13], SimCLR [6], SimSiam [7], and SwAV [5]. While SSL has transformed many computer vision applications, its potential to directly address shortcut learning in medical imaging remains underexplored.

In this study, we introduce ShortCXR, a shortcut-infused CXR dataset that systematically incorporates three types of biases—distortion, source, and demographic. We then investigate how state-of-the-art self-supervised learning (SSL) frameworks can mitigate these shortcuts and improve diagnostic outcomes. Through a series of four experiments, we measure the extent to which SSL can reduce bias and increase fairness. Our results underscore SSL's capacity to promote more reliable and equitable AI-driven diagnostics, offering practical guidance for designing models that are less prone to shortcuts and more robust in real-world clinical settings.

2 Method

2.1 Dataset

In this study, we utilized two chest X-ray (CXR) datasets. The COVID-19 Radiography Dataset (COVID) [8,20] includes data of COVID-19, as well as normal, lung opacity, and viral pneumonia images, comprising a total of 10,192 normal images, 3,067 COVID-19 images, 6,012 lung opacity images, and 1,348 viral pneumonia images. Additionally, we employed the MIMIC-CXR (MIMIC) dataset [12,16,17], a large, publicly available collection with 13 disease labels and three demographic labels: age, gender, and race. To ensure the integrity of the dataset division, all data was split into training, validation, and test sets in a 6:1:3 ratio, with a strict policy that individual subjects appear in only one set. We preprocessed the dataset to create a 4-class MIMIC dataset by filtering to retain cases with either normal findings or a single disease, selecting the four predominant conditions—Normal, Lung Opacity, Cardiomegaly, and Pleural Effusion, and balancing the dataset by downsampling the Normal cases. The final numbers of images in the MIMIC-CXR dataset are as follows: 2,192 for Normal, 2,147 for Lung Opacity, 1,566 for Cardiomegaly, and 1,910 for Pleural Effusion. Furthermore, the dataset was categorized into groups based on age (0–20, 20–40, 40–60, 60–80, and 80+), gender (Male and Female), and race (White, Black, and Asian) for fairness evaluation.

2.2 ShortCXR Dataset

2.2.1 Distortion-Related Shortcut

We evaluated all situations where shortcuts could potentially occur and defined four types: mark, contrast, lightness, and compression shortcuts. These were applied to the COVID dataset and the MIMIC dataset:

- (a) Mark: Four unique masks (− | = ||) were created and applied to the upper-left corner of the original images, introducing a distinct visual cue.
- (b) Contrast and Lightness: We adjusted the contrast and lightness of the images using four scales (0.6, 0.8, 1, 1.2), following a Gaussian distribution with $\sigma = 0.02$, to simulate variations in image exposure and contrast.
- (c) Compression. The original images underwent compression at four different quality levels (100%, 60%, 20%, 6%) in JPEG format, mimicking various degrees of image quality degradation that can occur in real-world scenarios.

We conducted experiments on datasets designed to examine distortion-related shortcuts, ensuring an equal number of categories for both target and shortcut labels, symbolized as $|A_t| = |A_b|$. We used a function $g : A_t \rightarrow A_b$ to create a one-to-one correspondence between target and shortcut labels, meaning each target label is paired with a unique shortcut label. For each shortcut label, we define the flipped shortcut label as the combination of the remaining three labels and construct flipped dataset. We borrowed the concepts of bias-align and bias-conflict from previous research [18], applying them to define shortcut-align and shortcut-conflict scenarios in our study. The proportion of shortcut-conflict samples was set at either 0.01 or 0.05.

2.2.2 Source-Related Shortcut

We analyzed the 'Lung Opacity' label across the MIMIC and COVID datasets, identifying differences in their distributions through the examination of average images and histograms. Consequently, we utilized all 'Lung Opacity' images from the COVID dataset, while sourcing the remaining images from MIMIC. This approach was predicated on the assumption that the model would more readily identify 'Lung Opacity' due to this targeted selection of higher-quality COVID dataset images.

2.2.3 Demographics-Related Shortcut

We analyzed the MIMIC dataset, focusing on gender, age, and race. Using the chi-squared test, we examined the link between demographic and disease labels, uncovering significant disparities across groups. All p-values were under 0.001, suggesting demographic information could lead to shortcuts in model predictions.

2.3 Experiment

The experimental setup is illustrated in Fig. 1. Our approach began with a baseline model featuring a ResNet18 architecture. We compared this baseline against four advanced SSL techniques: SimCLR, BYOL, SimSiam, and SwAV. Given that most SSL techniques incorporate data augmentation in their training process, we also compare the results with those of a baseline model employing similar augmentation settings. We evaluated four distinct augmentation strategies: CJB, JB, CB, and CJ, where 'C' represents cropping, 'J' color jittering, and 'B' blurring.

In our feature projection analysis, we applied the ResNet18 baseline to extract 512-dimensional features from the penultimate average pooling layer, preceding the final linear classifier. This procedure was uniformly applied in both the baseline and the linear classifier models, with the Area Under the ROC Curve (AUC) score on the validation set as the key metric for evaluation, regardless of the focus on disease labels or shortcut labels.

For evaluating Shortcut Information, we treated shortcut classification as a downstream task for SSL methodologies and compared their performance with SOTA SSL methods. Furthermore, we executed a comprehensive ablation study on augmentation techniques within the context of BYOL, providing a deeper insight into the influence of these methods on model performance.

Finally, we employed GradCAM [21] to visually highlight the specific areas of focus for the model. For the baseline model, GradCAM was applied to layer 4 of the ResNet backbone with the predicted label. In the case of SSL methods, we connected the SSL feature extractor, which also utilized the ResNet18 backbone, along with a downstream linear classifier. Subsequently, GradCAM was applied to layer 4 of the ResNet backbone with the predicted label.

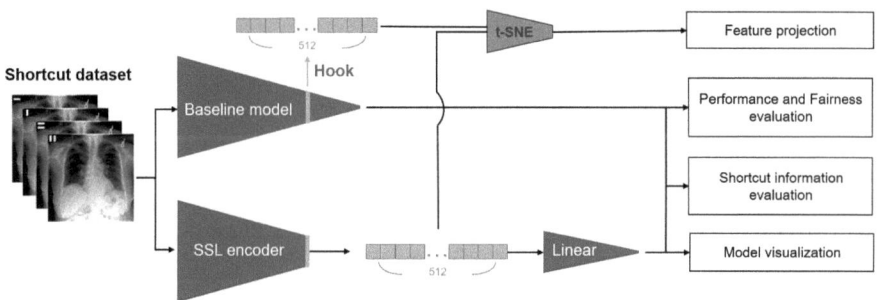

Fig. 1. Overview of Experiments. This figure summarizes the methodology for four distinct experiments designed to investigate the influence of shortcuts on model performance.

2.4 Fairness Evaluation

In binary classification, we define the input as X and an unknown target as Y. We observe the input X and make a prediction $\hat{Y} = f(X)$. The function f, which maps our input to our prediction $\hat{Y}$, is the classifier. In terms of fairness, we utilized the separation metric defined in [3]. For binary classifiers, separation is equivalent to requiring for all groups a and b the two constraints:

$$\mathbb{P}\{\hat{Y} = 1 \mid Y = 1, A = a\} = \mathbb{P}\{\hat{Y} = 1 \mid Y = 1, A = b\}, \tag{1}$$
$$\mathbb{P}\{\hat{Y} = 1 \mid Y = 0, A = a\} = \mathbb{P}\{\hat{Y} = 1 \mid Y = 0, A = b\}. \tag{2}$$

The true positive rate is the complement of the false negative rate. Consequently, separation requires that all groups experienced the same false negative rate and the same false positive rate, effectively demanding error rate parity. In cases with more than two groups, we aimed to equalize error rates. To achieve this, we computed the fairness across all possible group combinations and selected the maximum value. The function $fairness$ maps the groups to fairness, and we calculated fairness in three groups or more:

$$fairness(a, b, c) = max(fairness(a, b), fairness(b, c), fairness(a, c)). \tag{3}$$

For the distortion-related shortcut, we computed fairness by assessing the 4-class average error rate parity between the shortcut dataset (across all distortion-related shortcut types {mark, contrast, lightness, compression} and all conflict ratios {0.01, 0.05}) and the corresponding flipped shortcut dataset, investigating the impact of shortcuts. For the source-related shortcut, we calculated error parity between the Lung Opacity shortcut dataset and the original MIMIC dataset, exploring the influence of different data sources. For the demographics-related shortcut, we determined fairness for gender, age, and race.

3 Experimental Result

3.1 Performance and Fairness Evaluation

In the COVID distortion-related shortcut dataset (Fig. 2a), SimCLR exhibits commendable performance and enhanced fairness, with a lower score indicating superior fairness. The narrower the gap between the two settings (0.05 and 0.01), the greater the robustness of the method. The baseline model focuses primarily on shortcuts, resulting in both high performance and high fairness. Notably, we observe that across all distortion-related datasets, an increased conflict ratio makes it harder for the model to learn shortcuts, consequently enhancing fairness across all methods. In the MIMIC demographics-related shortcut dataset (Fig. 2b), SwAV stands out as the only method capable of surpassing the baseline in both performance and fairness, particularly in addressing gender-based shortcuts. In the MIMIC source-related shortcut dataset (Fig. 2c and 2d), we compare the results obtained from models trained on the Lung Opacity shortcut dataset and the Original dataset. We evaluate their performance using the Lung Opacity shortcut test set and assess fairness as defined in Sect. 2.4.

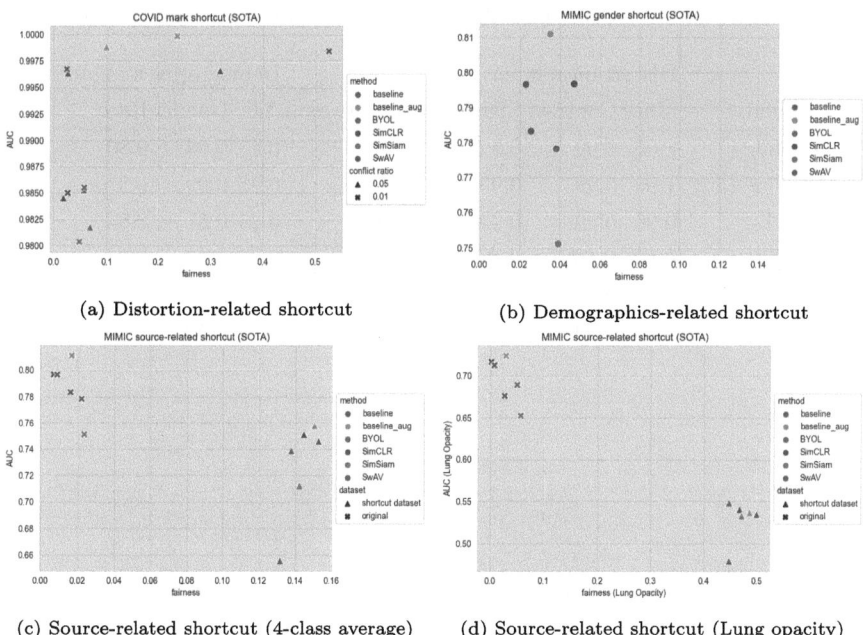

Fig. 2. Performance versus fairness for different shortcuts.

3.2 Shortcut Information Evaluation

As illustrated in Table 1, SimCLR helps reduce the model's dependence on distortion-related shortcut cues during disease identification. On the other hand, SwAV excels in both performance and fairness evaluations (Fig. 2b), as well as demographic information assessment. This indicates SwAV's adeptness in extracting representative features conducive to demographic classification, such as age, gender, or race. In contrast, SimSiam exhibits the worst performance in terms of demographic classification and disease diagnosis.

3.3 Feature Projection and Model Visualization

In distortion-related shortcuts, it is shown that baseline models easily pick up shortcuts, resulting in a distinct separation in the distribution. Conversely, the distribution achieved by SwAV demonstrates a reduced dependency on shortcuts, demonstrating its ability to mitigate the influence of such biases (Fig. 3).

In the original dataset, SwAV demonstrates a more focused attention on specific areas of the lung. Within the distortion-related dataset, SwAV's attention on the lung fields persists for COVID and Viral Pneumonia cases. However, for Normal and Lung Opacity cases, it exhibits a propensity to rely on shortcuts as indicators of pathology, as shown in Fig. 4a. This behavior aligns with the characteristics of SwAV discussed in Sect. 3.2. Regarding source-related shortcuts, the analysis of examples under the Lung Opacity label indicates that the

Table 1. AUCs for distortion-related and demographics-related shortcuts.

Method	Distortion-related				Demographics-related		
	Mark	Contrast	Lightness	Compression	Age	Gender	Race
baseline	1.00	0.99	0.99	0.99	0.85	0.99	0.86
baseline (aug)	1.00	0.99	0.99	1.00	0.88	0.99	0.87
BYOL	0.96	0.99	0.99	**0.96**	0.85	0.97	0.84
SimCLR	**0.94**	**0.97**	**0.96**	0.98	0.85	0.97	0.84
SimSiam	0.95	0.97	0.96	0.97	**0.80**	**0.88**	**0.72**
SwAV	0.98	0.99	0.99	1.00	0.86	0.99	0.85

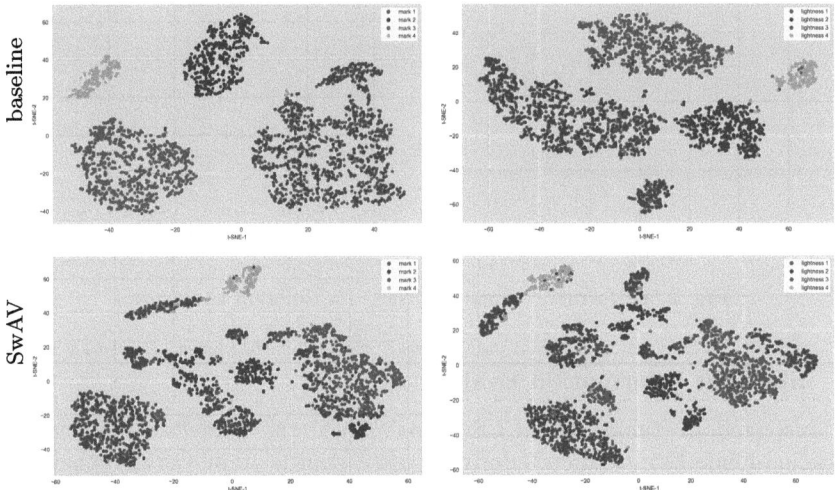

Fig. 3. Feature projection in distortion-related (Left: mark; Right: lightness) shortcut.

model is capable of concentrating on relevant pathological features. Across the remaining three classes, SwAV outperforms the baseline model, as illustrated in Fig. 4b. For example, SwAV utilized the de-identification mask to a lesser extent than the baseline model.

4 Discussion

We evaluated the performance and fairness of four SOTA SSL methods in comparison to a baseline ResNet model, across a variety of shortcut datasets, including those related to distortion, source, and demographics. Our findings indicate enhancements in both performance and fairness metrics across all examined shortcut datasets. Among the SSL methods, those models that demonstrated superior performance were also found to consistently generate relevant features that performed well in other classification tasks. Furthermore, we reveal how

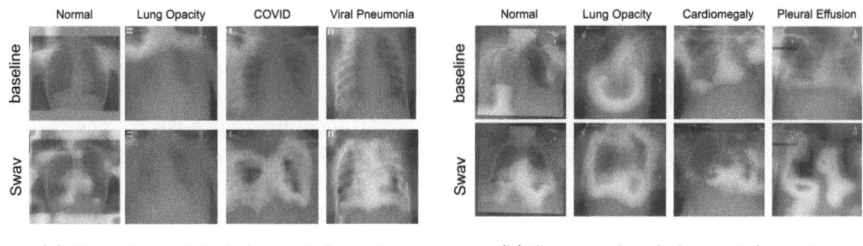

(a) Distortion-related shortcut dataset   (b) Source-related shortcut dataset

Fig. 4. Model visualization in shortcut dataset on baseline and SwAV

GradCAM visualizations identify the lung fields as primary areas of focus for models that exhibit both high performance and fairness, underscoring the potential of SSL methods in mitigating biases and enhancing diagnostic accuracy.

Based on our experimental findings with the four SSL methods, we hypothesize that cluster-based methods like SwAV retain all useful features, including both shortcut features and pathological features. Consequently, SwAV excels in addressing high-level shortcuts, such as demographics-related shortcuts. In contrast, BYOL and SimCLR tend to focus more on the most useful features, making them better suited for handling low-level shortcuts, including distortion-related and source-related shortcuts. However, it is worth noting that there exist numerous SSL methods based on alternative concepts, such as Barlow Twins [23] and VICReg [2], both rooted in feature decorrelation through contrastive learning. Additionally, methods based on masked image modeling, like BEiT [1] and MAE [15], offer alternative approaches worth considering.

In the context of augmentation strategies within SSL, the exclusion of color jittering tends to lead the model toward learning shortcuts associated with contrast and lightness. Similarly, omitting blurring encourages the model to adopt shortcuts related to compression features. Within demographics-related datasets and the MIMIC source-related shortcut dataset, the CJB augmentation strategy consistently surpasses other combinations in multiple dimensions of performance.

5 Conclusion

In this study, we outlined three shortcut types (distortion-related, source-related, demographics-related) and developed fairness evaluation methods to address them. Through experiments, we gained insights into the impact of SSL on model behavior and shortcut mitigation. The results affirm SSL's effectiveness in enhancing fairness and interpretability in models dealing with shortcut challenges.

Disclosure of Interests. The authors have no competing interests to declare.

References

1. Bao, H., Dong, L., Piao, S., Wei, F.: BEiT: BERT pre-training of image transformers. arXiv preprint arXiv:2106.08254 (2021)
2. Bardes, A., Ponce, J., LeCun, Y.: VICReg: variance-invariance-covariance regularization for self-supervised learning. arXiv preprint arXiv:2105.04906 (2021)
3. Barocas, S., Hardt, M., Narayanan, A.: Fairness and machine learning: limitations and opportunities. fairmlbook.org (2019). http://www.fairmlbook.org
4. Brown, A., Tomasev, N., Freyberg, J., Liu, Y., Karthikesalingam, A., Schrouff, J.: Detecting shortcut learning for fair medical AI using shortcut testing. Nat. Commun. **14**(1), 4314 (2023)
5. Caron, M., Misra, I., Mairal, J., Goyal, P., Bojanowski, P., Joulin, A.: Unsupervised learning of visual features by contrasting cluster assignments. In: Larochelle, H., Ranzato, M., Hadsell, R., Balcan, M., Lin, H. (eds.) Advances in Neural Information Processing Systems, vol. 33, pp. 9912–9924. Curran Associates, Inc. (2020)
6. Chen, T., Kornblith, S., Norouzi, M., Hinton, G.: A simple framework for contrastive learning of visual representations. In: Proceedings of the 37th International Conference on Machine Learning, ICML 2020. JMLR.org (2020)
7. Chen, X., He, K.: Exploring simple Siamese representation learning. In: Proceedings of the IEEE/CVF Conference on Computer Vision and Pattern Recognition (CVPR), pp. 15750–15758 (2021)
8. Chowdhury, M.E., et al.: Can AI help in screening viral and covid-19 pneumonia? IEEE Access **8**, 132665–132676 (2020)
9. DeGrave, A.J., Janizek, J.D., Lee, S.I.: AI for radiographic COVID-19 detection selects shortcuts over signal. Nat. Mach. Intell. **3**(7), 610–619 (2021). https://doi.org/10.1038/s42256-021-00338-7
10. Gichoya, J.W., et al.: AI recognition of patient race in medical imaging: a modelling study. Lancet Digit. Health **4**(6), e406–e414 (2022). https://doi.org/10.1016/S2589-7500(22)00063-2. https://www.ncbi.nlm.nih.gov/pmc/articles/PMC9650160/
11. Gidaris, S., Singh, P., Komodakis, N.: Unsupervised representation learning by predicting image rotations (2018)
12. Goldberger, A.L., et al.: Physiobank, physiotoolkit, and physionet. Circulation **101**(23), e215–e220 (2000). https://doi.org/10.1161/01.CIR.101.23.e215. https://www.ahajournals.org/doi/abs/10.1161/01.CIR.101.23.e215
13. Grill, J.B., et al.: Bootstrap your own latent a new approach to self-supervised learning. In: Proceedings of the 34th International Conference on Neural Information Processing Systems, NIPS 2020. Curran Associates Inc., Red Hook, NY, USA (2020)
14. Gui, J., Chen, T., Cao, Q., Sun, Z., Luo, H., Tao, D.: A survey of self-supervised learning from multiple perspectives: algorithms, theory, applications and future trends (2023). https://doi.org/10.48550/arXiv.2301.05712
15. He, K., Chen, X., Xie, S., Li, Y., Dollár, P., Girshick, R.: Masked autoencoders are scalable vision learners. In: Proceedings of the IEEE/CVF Conference on Computer Vision and Pattern Recognition, pp. 16000–16009 (2022)
16. Johnson, A.E.W., et al.: MIMIC-CXR, a de-identified publicly available database of chest radiographs with free-text reports. Sci. Data **6**(1), 317 (2019). https://doi.org/10.1038/s41597-019-0322-0
17. Johnson, A.E.W., Pollard, T.J., Berkowitz, S.J., Mark, R.G., Horng, S.: MIMIC-CXR database (version2.0.0). PhysioNet (2019). https://doi.org/10.13026/C2JT1Q

18. Nam, J., Cha, H., Ahn, S., Lee, J., Shin, J.: Learning from failure: De-biasing classifier from biased classifier. In: Advances in Neural Information Processing Systems, vol. 33, pp. 20673–20684 (2020)
19. Noroozi, M., Favaro, P.: Unsupervised learning of visual representations by solving Jigsaw puzzles (2017)
20. Rahman, T., et al.: Exploring the effect of image enhancement techniques on COVID-19 detection using chest X-ray images. Comput. Biol. Med. **132**, 104319 (2021). https://doi.org/10.1016/j.compbiomed.2021.104319. https://www.sciencedirect.com/science/article/pii/S001048252100113X
21. Selvaraju, R.R., Cogswell, M., Das, A., Vedantam, R., Parikh, D., Batra, D.: GRAD-CAM: visual explanations from deep networks via gradient-based localization. In: Proceedings of the IEEE International Conference on Computer Vision, pp. 618–626 (2017)
22. Wang, R., Kuo, P.C., Chen, L.C., Seastedt, K.P., Gichoya, J.W., Celi, L.A.: Drop the shortcuts: image augmentation improves fairness and decreases AI detection of race and other demographics from medical images. EBioMedicine **102** (2024)
23. Zbontar, J., Jing, L., Misra, I., LeCun, Y., Deny, S.: Barlow twins: self-supervised learning via redundancy reduction. In: International Conference on Machine Learning, pp. 12310–12320. PMLR (2021)
24. Zhu, J.Y., Park, T., Isola, P., Efros, A.A.: Unpaired image-to-image translation using cycle-consistent adversarial networks. In: 2017 IEEE International Conference on Computer Vision (ICCV), pp. 2242–2251 (2017). https://doi.org/10.1109/ICCV.2017.244

How Fair are Foundation Models? Exploring the Role of Covariate Bias in Histopathology

Abubakr Shafique[1](✉), Amanda Dy[1], Xiaoli Qin[1], Najd Alshamlan[2], Susan J. Done[3,4], Dimitrios Androutsos[1], and April Khademi[1,5,6,7,8]

[1] Department of Electrical, Computer and Biomedical Engineering, Toronto Metropolitan University, Toronto, ON, Canada
{abubakr.shafique,akhademi}@torontomu.ca
[2] King Abdullah bin Abdulaziz University Hospital, Princess Norah University, Riyadh, Saudi Arabia
[3] Laboratory Medicine Program, University Health Network, Toronto, ON, Canada
[4] Department of Laboratory Medicine and Pathobiology, University of Toronto, Toronto, ON, Canada
[5] Institute for Biomedical Engineering, Science Tech (iBEST), Toronto, ON, Canada
[6] Vector Institute for Artificial Intelligence, Toronto, ON, Canada
[7] Department of Medical Imaging, University of Toronto, Toronto, ON, Canada
[8] Institute of Medical Science, University of Toronto, Toronto, ON, Canada

Abstract. Foundation models (FMs) have introduced new opportunities for zero-shot generalization in downstream tasks through techniques such as linear probing and feature extraction. However, a systematic evaluation of their fairness regarding their sensitivity to covariate bias remains lacking, which is critical for clinical translation. In this study, we address this gap by constructing a unique dataset comprising identical glass slides digitized using two scanners, enabling a controlled simulation of covariate bias in data distributions. The same tissue regions (patches) are extracted from Whole Slide Images acquired on each scanner and processed through FMs to obtain zero-shot feature representations. We define and quantify 'representation shift', as the difference in feature vectors across scanners, and assess using metrics such as mean squared error, KullbackLeibler divergence, and a novel clustering-based Calinski-Harabasz index. Our results demonstrate that FMs exhibit significant scanner-dependent variability in feature representations, highlighting a key generalization limitation. This sensitivity to acquisition device introduces the risk of unequal or unfair performance; an important consideration for the safe and fair deployment of FMs in clinical settings.

Keywords: Covariate Bias · Representation Shift · Foundation Model · Domain Generalization · Fairness in Medical Imaging

1 Introduction

Medical Artificial Intelligence (AI) offers substantial improvements in diagnostics, therapeutic decision-making, and clinical workflow optimization [20]. However, despite the promise, critical challenges remain related to algorithm fairness and equity in healthcare [4,8]. Algorithm fairness relates to whether an AI system provides equitable performance across diverse data distributions [25]. In histopathology, these biases, characterized as domain shifts, are related to differences between the training and testing dataset, including Covariate Shift, Prior Shift, Posterior Shift, and Class-Conditional Shift [13]. Covariate Shift, which is bias that originates from differences in scanning devices across institutions, presents the leading barrier for wide-scale adoption and robust deployment of computational pathology systems across labs [11,15]. Such shifts can significantly impair the model's generalizability and effectiveness on Out-Of-Distribution (OOD) samples [15]. To enhance generalizability and provide a flexible framework for a wide range of zero-shot downstream tasks, foundation models (FMs) have recently been introduced in histopathology [3].

The term *"Foundation Model"* refers to large-scale, deep neural networks that have been extensively pre-trained on massive datasets, enabling a versatile core architecture for a broad range of downstream tasks and applications [17,26]. FMs have been positioned as the solution for domain shift, as the models are developed from large datasets and can learn generalizable representations, which holds great promise for accelerating translation opportunities. FMs are rapidly transforming fields such as Natural Language Processing (NLP) [9], Computer Vision [16], Computational Pathology [1,7,12,24,27], and beyond [14,29]. Ideally, a robust foundation model should be invariant to domain-specific variations, but FMs have not yet been evaluated comprehensively in terms of fairness, generalization, and robustness to the covariate bias, common in real-world scenarios.

In this paper, we interpret fairness as the robustness of a model to covariate shifts introduced by different scanners, which can result in unequal performance in healthcare settings. This is distinct from fairness definitions involving protected attributes such as race or gender, and instead focuses on infrastructural bias that may compromise model generalizability. We quantify the covariate shift from the zero-shot feature maps from FMs due to different scanning devices using a dataset of tissue slides acquired by two different scanners. This dataset offers an ideal setting in which all other variables are controlled, allowing the impact of scanning devices to be analyzed. We aim to study whether FMs are fair through resilience to scanner-related variability by measuring differences in the representations.

1.1 Foundation Models in Histopathology

Advances in Self-Supervised Learning (SSL) have facilitated the training of FMs by enabling the extraction of meaningful representations from unlabeled data [5,7,19,24,27]. As a result, numerous FMs have been developed, leveraging SSL techniques. SSL addresses the challenge of data inefficiency by deriving

Table 1. Histopathology DL models available in the literature.

Models	Params. (M)	Architecture	Patches (M)	WSIs (K)	Patch Size	Vector Dim.	Data	Organs
KimiaNet [21]	7	DenseNet-121	0.3	11	1000	1024	TCGA[2]	30
PathDino [2]	9	ViT-S/16[1]	6	12	512	384	TCGA[2]	33
HIPT [6]	21	ViT-S/16	104	11	224	384	TCGA[2]	33
iBOT-Path [10]	85	ViT-B/16	40	6	224	768	TCGA[2]	13
Hibou-B [18]	85	ViT-B/14	512	1,139	224	768	Private	12
UNI [7]	303	ViT-L/16	100	100	224	1024	MGB[3]	20
Virchow [24]	632	ViT-H/14	2,000	1,488	224	2560	MSKCC[4]	17
Virchow2 [30]	632	ViT-H/14	2,000	3,100	224	2560	MSKCC[4]	200
GigaPath [27]	1,135	ViT-G/14	1,385	171	224	1536	PHS[5]	31

[1]PathDino is a modified ViT-S/16 with only 5 blocks. [2]TCGA: The Cancer Genome Atlas. [3]MGB: Mass General Brigham. [4]MSKCC: Memorial Sloan Kettering Cancer Center. [5]PHS: Providence Health and Services.

supervisory signals intrinsically from the data, eliminating the dependence on manually annotated labels [5,19]. Vast amounts of unlabeled data are instead used to learn generalizable feature representations that can be efficiently adapted to various downstream tasks [5,19]. It is especially advantageous in digital pathology (DP), where labelled data is scarce [2,23].

Recently, several histopathology FMs have been introduced, including UNI [7], Hibou-B [18], Virchow [24], Virchow2/2G [30], and GigaPath [27]. These models have adopted the DINOv2 SSL framework [19], capitalizing on large-scale proprietary datasets. Several other FMs trained on large-scale histopathology datasets have been proposed, such as iBOT-Path [10], which leverages the iBOT self-supervised learning framework [28]. Similarly, FMs such as the Hierarchical Image Pyramid Transformer (HIPT) [6] and PathDino [2] use the original DINO framework [5] to enable representation learning in histopathology. While these are Vision Transformer (ViT)-based, there are also convolutional neural network (CNN) models, such as KimiaNet [21], which is built on the DenseNet-121 architecture and trained in a supervised manner using the full TCGA dataset. Table 1 summarizes the key characteristics of the considered FMs.

2 Materials and Methods

A novel framework is proposed for measuring the scanner-induced covariate shift in histopathology FMs. A fair and generalizable FM should produce scanner-invariant representations when analyzing the same tissue sample imaged across different devices. Therefore, we evaluate fairness as the degree to which FMs yield consistent representations under controlled, biologically matched but scanner-varied input. This can further understanding of algorithm fairness, bias, robustness and reliability in diverse image acquisition settings.

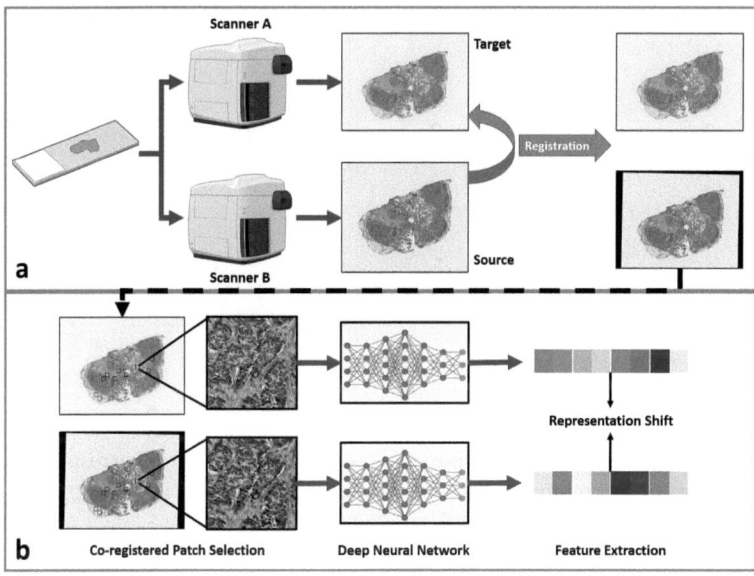

Fig. 1. Representation Shift. Quantifying the covariate bias (representation) shift of FMs. (a) a single glass slide is digitized using two distinct scanners, and the resulting images are subsequently aligned through an affine registration. Scanner A corresponds to the Aperio AT2, while Scanner B refers to the Sakura VisionTek scanner. (b) Co-registered patches are extracted from spatially aligned images, and the patches are processed through a FM to obtain feature representations. Extracted features are utilized to quantify the covariate bias and representation shift.

2.1 Representation Shift

Covariate bias in histopathology is related to the shift in data distributions from different scanning platforms, which cause discrepancies in feature embeddings, reduces generalizability and ultimately creates inequities for patients. In this work, we quantify scanner bias by measuring the representation shift of feature vectors produced by the same (zero-shot) FMs on identical tissue regions imaged using various devices; to our best knowledge, this is the first work of its kind.

To quantify the susceptibility of FMs (Table 1) to scanner bias, the feature representations of the same tissue slides are compared across scanners using three metrics: Mean Squared Error (MSE), KullbackLeibler (KL) divergence, and Calinski-Harabasz (CH). The dataset is defined as $\mathcal{D} = \{(x_i^A, x_i^B) \mid i = 1, 2, \ldots, N\}$, where $x_i^A \in \mathbb{R}^{H \times W \times C}$ denotes the i-th patch from Scanner A (Aperio), $x_i^B \in \mathbb{R}^{H \times W \times C}$ denotes the corresponding spatially aligned patch from Scanner B (VisionTek), and H, W, and C is the height, width, and channels per patch. Let $f_\theta : \mathbb{R}^{H \times W \times C} \to \mathbb{R}^d$ be a FM parameterized by θ, which maps an input patch to a d-dimensional feature space. The corresponding feature embeddings are obtained $z_i^A = f_\theta(x_i^A)$ and $z_i^B = f_\theta(x_i^B)$, Where $z_i^A, z_i^B \in \mathbb{R}^d$

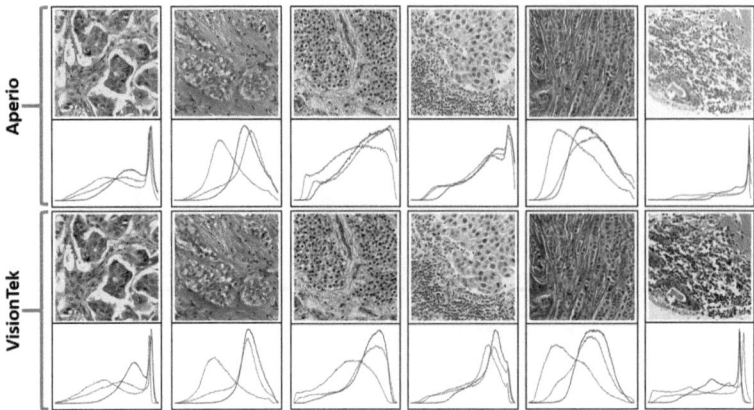

Fig. 2. Covariate Bias. Sample co-registered patches and their corresponding RGB histograms are shown to illustrate scanner-induced covariate bias. Histograms for each co-registered patch are presented on the same scale.

represent feature vectors extracted from the same tissue region scanned by Scanner A and Scanner B, respectively. To enable fair comparison across FMs with varying feature activation scales, the embeddings z_i^A and z_i^B are z-score normalized using the sample-specific mean and standard deviation, resulting in $\tilde{z}_i^A$ and $\tilde{z}_i^B$. These normalized embeddings are then used to compute the MSE, denoted as $MSE(\tilde{z}_i^A, \tilde{z}_i^B)$. Additionally, to place the feature embeddings on a common scale for KL divergence computation, Principal Component Analysis (PCA) is applied to reduce their dimensionality to 384, yielding $\hat{z}_i^A$ and $\hat{z}_i^B$. The KL divergence, denoted as $KL(\hat{z}_i^A, \hat{z}_i^B)$. Both MSE and KL divergence assess the similarity between paired feature vectors extracted from identical tissue regions and are sensitive to subtle shifts in representation. Notably, KL divergence operates on the probability distributions derived from the feature embeddings, providing a complementary perspective to MSE.

We also propose a novel use of the CH, $CH(\tilde{z}_i^A, \tilde{z}_i^B)$, clustering index [23] to measure the ratio of inter-cluster separation (distance between cluster centres) and intra-cluster dispersion (compactness of each cluster) of the Scanner A and Scanner B latent space using normalized embeddings. Lower CH indicates that embeddings from both scanners overlap with less bias.

2.2 Data Preparation

Two breast cancer datasets were created for this work: H&E-stained WSIs acquired from Ontario Institute for Cancer Research (OICR), Toronto, ON, Canada, and H&E-stained tissue microarrays (TMAs) acquired from University Health Network (UHN), Toronto, ON, Canada. The OICR dataset includes 111 formalin-fixed paraffin-embedded (FFPE) glass slides. Each slide was digitized using the Aperio AT2 at 40x magnification with 0.25 microns per pixel (mpp),

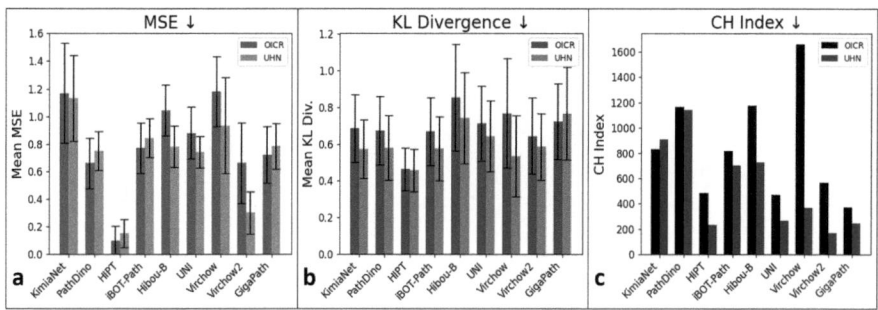

Fig. 3. Representation Shift. Representation shift across all SOTA FMs (see Table 1), quantified using (a) mean MSE, (b) mean KL divergence, and (c) CH Index for both the OICR and UHN datasets.

and the Sakura VisionTek at 20x magnification with 0.27 mpp, upsampled to match the resolution of the Aperio AT2 scans. The UHN dataset has a total of 559 TMA cores that were scanned with the Aperio ScanScope at 40x magnification with 0.25 mpp, and the Sakura VisionTek at 20x magnification with 0.27 mpp, again upsampled to match the resolution of the Aperio Scanscope scans. Samples were upsampled to 0.25 MPP for better spatial alignment.

To achieve spatial alignment of images across scanners, the WSIs and TMAs are aligned using affine registration [22] with the Aperio scans as the fixed reference and the VisionTek scans as the moving image, as depicted in Fig. 1(a). Co-registered tissue patches measuring 512×512 pixels were extracted at 0.5 mpp from the corresponding locations across scanners. A total of 4,952 patches for the OICR dataset were extracted per scanner, yielding a combined 9,904 patches for evaluation. A total of 2,041 identical tissue patches were extracted from the UHN dataset, yielding a total of 4,082 patches. Figure 2 presents sample images from both scanners, showing similar tissue morphology but noticeable color variations in their corresponding histograms due to scanner differences. While these slides may appear identical visually, such subtle color shifts can lead to biased outcomes in deep learning-based diagnosis.

3 Results and Discussion

The robustness and fairness of histopathology FMs (ranging 7M to 1.1B prameters) were evaluated through comprehensive experiments looking at the consistency of model-generated representations for similar tissue morphology from two different scanners. Figure 3 (a) and (b) shows the mean and standard deviations of MSE and KL divergence between registered tissue patches and Fig. 3 (c) displays the CH index. To analyze representation shifts between datasets we report the absolute difference in each metric shown in Fig. 4 (a) (c). The difference analysis highlights the consistency of FM behavior across datasets, indicating which models are more stable than others. While metrics like MSE

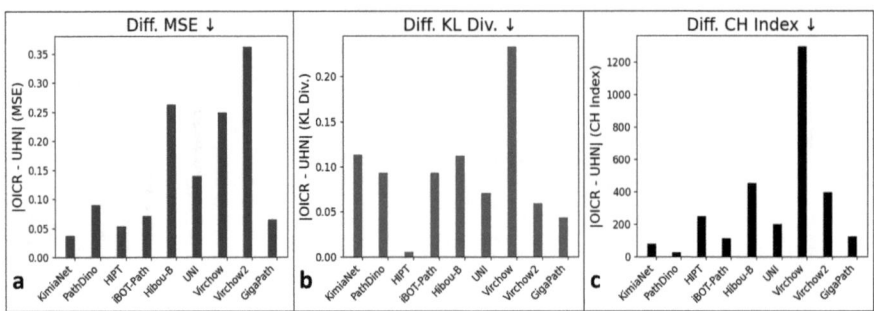

Fig. 4. Representation Shift across Datasets. Difference in representation shift of FMs between the OICR and UHN datasets, measured by (a) the absolute difference in mean MSE, (b) the absolute difference in mean KL divergence, and (c) the absolute difference in CH Index.

and KL divergence show only subtle variations, they help identify models with higher or lower consistency across datasets. Moreover, Fig. 5 presents the 2D t-SNE projections of the feature embeddings across datasets.

HIPT produces the most consistent representations across both datasets, achieving the lowest MSE and KL divergence. However, it does not yield the lowest CH index. GigaPath achieves the lowest CH index on the OICR dataset, while Virchow2 has the lowest CH index on the UHN dataset. The t-SNE projections in Fig. 5 further support these findings: HIPT's embeddings show substantial overlap across scanners but exhibit poor clustering structure. In contrast, GigaPath demonstrates clear clustering patterns in the OICR dataset despite overlap, while Virchow2 displays a moderate clustering structure in the UHN dataset. These metrics collectively suggest that no single FM is universally superior in all criteria. While some models excel in minimizing representation shift (e.g., HIPT), others provide better inter-scanner separability or clustering consistency, indicating that performance varies with dataset characteristics and the chosen evaluation metric.

Other than HIPT, PathDino yields the second lowest MSE on the OICR dataset, closely followed by Virchow2. On the UHN dataset, Virchow2 achieves the second lowest MSE and demonstrates improved consistency in feature representations compared to its predecessor, Virchow. This improvement may be attributed to the larger and more diverse training dataset used for Virchow2, as well as its multi-resolution ViT architecture, which is likely to have learned more generalized local and global features.

The FMs that exhibit the highest embedding overlap in the t-SNE projections across both datasets are UNI and HIPT, while UNI demonstrate relatively higher MSE and KL divergence with lower CH index. On the other hand, GigaPath and Virchow2 show notable feature overlap primarily in the UHN and OICR datasets, respectively, suggesting increased variability across datasets and potential lim-

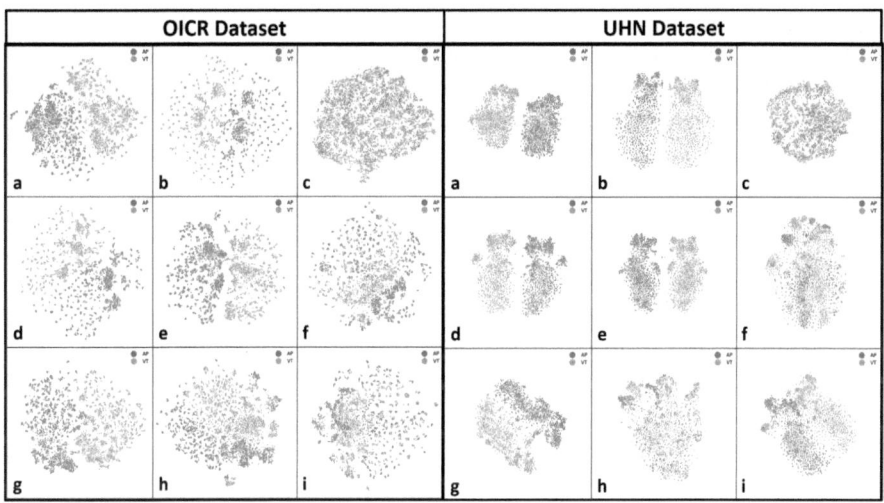

Fig. 5. t-SNE Projection. Visualization of feature embeddings for identical tissue patches using FMs in Table 1, including: (a) KimiaNet, (b) PathDino, (c) HIPT, (d) iBOT-Path, (e) Hibou-B, (f) UNI, (g) Virchow, (h) Virchow2, and (i) GigaPath. Red and blue points indicate embeddings from Aperio and VisionTek scanners, respectively.

itations in fairness and generalizability. As illustrated in Fig. 4 (a) - (c), KimiaNet, HIPT, and PathDino demonstrate the lowest variability across datasets for MSE, KL divergence, and CH index, respectively; however, they consistently yield relatively high values for MSE, KL divergence, and CH index, respectively. In Fig. 3 (a) Virchow shows the highest MSE on the OICR dataset, followed by KimiaNet. On the UHN dataset, KimiaNet exhibits the highest MSE, with Virchow ranking second highest. In terms of KL divergence, Hibou-B produces the highest values on both datasets, indicating a significant feature divergence between scanners. These results show that while some models maintain stable feature representations, they may still exhibit high inter-scanner discrepancy in terms of alignment and separation, highlighting the multifaceted nature of evaluating FM fairness and generalization. It is commonly assumed that large FMs trained on millions of WSIs inherently generalize well. However, our findings suggest that generalization performance is not solely determined by model size or dataset scale. For example, HIPT, a relatively smaller model, demonstrates consistent representations across different scanners (see Fig. 3). Similarly, UNI and Virchow2, though larger than HIPT but smaller than GigaPath, achieve competitive performance by producing comparable representations for similar tissue morphologies across scanners and datasets. In contrast, Virchow and Virchow2 share identical architectures and parameter counts, yet Virchow2 outperforms its predecessor in terms of MSE and CH index, while showing comparable performance in KL divergence. GigaPath, the largest model evaluated, performs comparably to significantly smaller models such as UNI and PathDino, suggesting diminishing returns with increased model scale. Interestingly, larger models like GigaPath, UNI, and Virchow2, as well as the smaller HIPT, all exhibit relatively low CH index values, despite their differences in size. This indicates that factors beyond model scale, such as training strategy and dataset composition, play a critical role in representation consistency and generalization.

4 Conclusion

Healthcare disparities in AI can arise from variations in data acquisition across clinical sites, such as differences in scanners, staining, and imaging protocols, leading to inconsistent model performance and reduced fairness. While FMs in histopathology aim to generalize across tasks using large-scale training data, their sensitivity to scanner-induced covariate shifts remains unclear. In this study, We isolate and assess the fairness of FMs to scanner-related bias by analyzing feature embeddings from co-registered tissue regions scanned with two different devices. A robust model should yield similar representations across scanners for similar tissue morphologies, which we evaluate using MSE, KL divergence, and the CH index. The results show that despite FMs being large models trained on lots of data, there is still a noticeable bias to image representation across scanners and datasets for many FMs.

No single model exhibits robust performance across all metrics. While HIPT shows low MSE and KL divergence, it has a high CH index. In contrast, UNI, GigaPath, and Virchow2 exhibit higher MSE but relatively lower or comparable CH index values, indicating better clustering. These findings are rooted in representation discrepancies observed for similar tissue morphologies scanned by different scanners, highlighting the influence of domain-specific acquisition protocols. Future downstream tasks, such as classification, may offer deeper insights into the generalization and potential biases of FMs. Additionally, studying variations in staining protocols could provide further understanding. Improving fairness and generalizability in histopathology AI requires training on diverse, multi-institutional datasets, standardized imaging protocols, and thorough evaluation to ensure equitable and reliable clinical deployment.

References

1. Alfasly, S., et al.: Foundation models for histopathology–fanfare or flair. Mayo Clin. Proc. Digit. Health **2**(1), 165–174 (2024)
2. Alfasly, S., Shafique, A., Nejat, P., Khan, J., Alsaafin, A., Alabtah, G.: Rotation-agnostic image representation learning for digital pathology. In: Proceedings of the IEEE/CVF Conference on Computer Vision and Pattern Recognition, pp. 11683–11693 (2024)
3. Awais, M., et al.: Foundation models defining a new era in vision: a survey and outlook. IEEE Trans. Pattern Anal. Mach. Intell. (2025)
4. Carey, S., Pang, A., de Kamps, M.: Fairness in AI for healthcare. Future Healthc. J. **11**(3), 100177 (2024)

5. Caron, M., et al.: Emerging properties in self-supervised vision transformers. In: Proceedings of the IEEE/CVF International Conference on Computer Vision, pp. 9650–9660 (2021)
6. Chen, R.J., et al.: Scaling vision transformers to gigapixel images via hierarchical self-supervised learning. In: Proceedings of the IEEE/CVF Conference on Computer Vision and Pattern Recognition, pp. 16144–16155 (2022)
7. Chen, R.J., et al.: Towards a general-purpose foundation model for computational pathology. Nat. Med. **30**(3), 850–862 (2024)
8. Chen, R.J., et al.: Algorithmic fairness in artificial intelligence for medicine and healthcare. Nat. Biomed. Eng. **7**(6), 719–742 (2023)
9. Chowdhery, A., et al.: PaLM: scaling language modeling with pathways. J. Mach. Learn. Res. **24**(240), 1–113 (2023)
10. Filiot, A., et al.: Scaling self-supervised learning for histopathology with masked image modeling. medRxiv (2023)
11. Howard, F.M., et al.: The impact of site-specific digital histology signatures on deep learning model accuracy and bias. Nat. Commun. **12**(1), 4423 (2021)
12. Huang, Z., Bianchi, F., Yuksekgonul, M., Montine, T.J., Zou, J.: A visual-language foundation model for pathology image analysis using medical twitter. Nat. Med. **29**(9), 2307–2316 (2023)
13. Jahanifar, M., et al.: Domain generalization in computational pathology: survey and guidelines. ACM Comput. Surv. (2023)
14. Khanna, S., et al.: DiffusionSat: a generative foundation model for satellite imagery. In: The Twelfth International Conference on Learning Representations (2023)
15. Kheiri, F., Rahnamayan, S., Makrehchi, M., Asilian Bidgoli, A.: Investigation on potential bias factors in histopathology datasets. Sci. Rep. **15**(1), 11349 (2025)
16. Kirillov, A., et al.: Segment anything. In: Proceedings of the IEEE/CVF International Conference on Computer Vision, pp. 4015–4026 (2023)
17. Moor, M., et al.: Foundation models for generalist medical artificial intelligence. Nature **616**(7956), 259–265 (2023)
18. Nechaev, D., Pchelnikov, A., Ivanova, E.: Hibou: a family of foundational vision transformers for pathology. arXiv preprint arXiv:2406.05074 (2024)
19. Oquab, M., et al.: DINOv2: learning robust visual features without supervision. arXiv preprint arXiv:2304.07193 (2023)
20. Rajpurkar, P., Chen, E., Banerjee, O., Topol, E.J.: AI in health and medicine. Nat. Med. **28**(1), 31–38 (2022)
21. Riasatian, A., et al.: Fine-tuning and training of DenseNet for histopathology image representation using TCGA diagnostic slides. Med. Image Anal. **70**, 102032 (2021)
22. Shafique, A., Babaie, M., Sajadi, M., Batten, A., Skdar, S.: Automatic multi-stain registration of whole slide images in histopathology. In: 2021 43rd Annual International Conference of the IEEE Engineering in Medicine & Biology Society (EMBC), pp. 3622–3625. IEEE (2021)
23. Shafique, A., et al.: A preliminary investigation into search and matching for tumor discrimination in world health organization breast taxonomy using deep networks. Mod. Pathol. **37**(2), 100381 (2024)
24. Vorontsov, E., et al.: A foundation model for clinical-grade computational pathology and rare cancers detection. Nat. Med., 1–12 (2024)
25. Wang, Y., Song, Y., Ma, Z., Han, X.: Multidisciplinary considerations of fairness in medical AI: a scoping review. Int. J. Med. Inf. **178**, 105175 (2023)
26. Wei, J., et al.: Emergent abilities of large language models. arXiv preprint arXiv:2206.07682 (2022)

27. Xu, H., et al.: A whole-slide foundation model for digital pathology from real-world data. Nature, 1–8 (2024)
28. Zhou, J., et al.: iBOT: image BERT pre-training with online tokenizer. arXiv preprint arXiv:2111.07832 (2021)
29. Zhou, Y., et al.: A foundation model for generalizable disease detection from retinal images. Nature **622**(7981), 156–163 (2023)
30. Zimmermann, E., et al.: Virchow 2: scaling self-supervised mixed magnification models in pathology. arXiv preprint arXiv:2408.00738 (2024)

The Cervix in Context: Bias Assessment in Preterm Birth Prediction

Joris Fournel[1](✉), Paraskevas Pegios[1], Emilie Pi Fogtmann Sejer[2], Martin Tolsgaard[2], and Aasa Feragen[1]

[1] Technical University of Denmark, Kongens Lyngby, Denmark
jorisfournell@gmail.com
[2] Region Hovedstaden Hospital, Copenhagen, Denmark

Abstract. The death and disability burden of spontaneous preterm births (sPTB) strikes millions of infants every year. Early risk stratification using transvaginal ultrasound imaging allows clinicians to implement preventive measures to delay or avoid preterm delivery. This makes sPTB prediction a highly valuable clinical target. Machine learning (ML) models have outperformed the clinical baseline—namely, cervical length (CL)—in numerous retrospective studies. Yet in practice, sPTB prevention still relies on the modest 40% sensitivity achieved by the standard CL threshold. A major barrier to clinical adoption of ML-based predictors is the lack of bias assessment. ML performance is known to vary across subpopulations, but studies rarely quantified where and for whom ML actually improves upon CL. Without subgroup-specific performance insights, clinicians are unable to identify the patients for whom ML would help—or harm—leading to justified reluctance toward its deployment.

In our analysis, CL suffered its largest performance drops across physiological and patient-related parameters, whereas the deep learning model had a stronger sensitivity to imaging-related attributes.

As a byproduct, we report a promising finding: Ultrasound images that display the cervix's wide anatomical context appears to strongly boost deep learning performance.

Keywords: Preterm Birth · Deep Learning · Bias Assessment

1 Introduction

One out of ten infants enters the world prematurely, facing increased risk of neonatal death, infant death, and future handicap [2,14,16]. Early screening enables preventive measures such as progesterone administration and cervical cerclage, and can thus prevent or delay spontaneous preterm birth (sPTB) [8,10]. The current gold standard for screening – thresholding the cervical length (CL) as measured in a transvaginal ultrasound scan – leaves room for improvement: A sensitivity located between 30 and 40% indicates a failure to detect the majority of high risk pregnancies [6,12].

Looking beyond CL, researchers have hypothesized the existence of more discriminative information within the cervical scan. Promising results have been found with machine learning-based texture analysis, initially reported on small datasets with less than 50 preterm birth cases [3–5,15,19]. Pegios et al. [17] showed, on a dataset with several thousands preterm births, that a deep learning classifier could raise the AUC level from 0.67 to 0.75.

However, no newborn has yet benefited from these alleged improvement in risk stratification. Why? None of these studies have contextualized the performance of their algorithms by providing a thorough analysis of model bias. The propensity of machine learning models to exhibit reduced performance across specific demographics and imaging domains is now well recognized within the scientific community, including in ultrasound imaging [7]. Only a stratified evaluation of performance across all relevant subgroups can unveil those disparities. Without it, bias opacity remains a major barrier to clinical adoption, as doctors, in fear of ML performance drop on the specific input at hand, will choose to rely on CL. Obstetricians need clear guidance on when to utilize automated risk scores, and when to fall back on traditional CL-based assessments.

We Aim to Bridge that Gap by Elucidating the Biases in Existing Models, Moving One Step Closer to Preventing More Preterm Births in Clinical Practice. We perform a comprehensive bias analysis of the three leading approaches to sPTB prediction: CL, texture analysis, and deep learning classifiers. While certain factors, such as demographics, are immutable, others—like imaging protocols and exam scheduling—can be adjusted. A secondary objective is to identify, through this bias analysis, the optimal configuration that maximizes predictive performance for preterm birth.

Our Key Contributions are as Follows:

1. A novel analysis: The first bias assessment across three different types of models for spontaneous preterm birth prediction.
2. A clinical answer: When is deep learning superior to CL, for which patients, and why?

2 Methods

Dataset. Our dataset comprises 7862 transvaginal ultrasound (US) images extracted from a national fetal ultrasound screening database [17]. Exclusion criteria included multiple pregnancy, placenta previa, labor induction, cesarean section, and gestational age (GA) <12+0 weeks or >33+6 weeks at the time of examination. The dataset was balanced, with an equal number of term (n = 3931) and preterm (n = 3931) birth images. Available clinical information included maternal characteristics (ethnicity, age, body mass index (BMI), parity, and cervical length), fetal parameters (gestational age, birth weight), imaging features (device name, pixel spacing), and environmental factors (place and year of birth).

Models. We evaluated three predictive approaches: cervical length (CL), the clinical gold standard for sPTB screening; a texture-based radiomics classifier, which has been shown in multiple studies to outperform CL [4,5,9]; and a deep learning (DL) model, previously demonstrated to surpass both CL and texture-based methods on a substantially larger dataset [17]:

1. **Cervical length (CL).** Manual CL measurements, recorded directly on the ultrasound images by sonographers, were extracted using DocTR, an open-source optical character recognition system [13]. CL was successfully extracted from 6194 of the 7862 images.
2. A **radiomics classifier** was constructed following best practices from the literature. Feature extraction included local binary patterns (Scikit-image [18]) and radiomic features (PyRadiomics [11]) across four regions of interest: The full cervix, the full anterior lip, and the outer portions of the anterior and posterior lips. Curvilinear masking of the cervix was recovered from SA-SonoNet segmentation head [17]. Broken lines were (eventually) reconnected before landmark detection and connection, region fill, resampling to common spacing and ROI-based Z-score normalization. A logistic regression model was trained on the concatenated features to predict sPTB.
3. A **deep learning classifier (DL)**; namely SA-SonoNet, a shape- and spatially-aware model described in Pegios et al. [17]. This model predicts sPTB from ultrasound images and their pixel spacing metadata.

For the DL and radiomics models, predicted preterm probabilities were generated for all images using 10-fold cross-validation. When comparing automated models to the CL baseline, the bias analysis was restricted to the 6194 images with available CL. 100-fold bootstrap scoring and MannWhitney U test was used to assess statistical significance between methods.

Bias Assessment Procedure. Model bias was assessed per attribute by analyzing whether classification performance significantly differed across its subgroups. Attributes with more than 10 unique values (e.g., maternal age, BMI, GA at birth, birth weight, birth year, and pixel spacing) were discretized into five quantile-based subgroups. Categorical variables were used as-is. For each subgroup, we computed performance metrics relevant for preterm birth screening: the sensitivity at 85% specificity, and the area under the receiver operating characteristic curve (AUC-ROC). (These metrics (contrarily to e.g. accuracy) are invariant to class prevalence.) A large performance gap between subgroups indicated potential bias associated with the corresponding attribute.

3 Results

3.1 Bias Detection in CL, Radiomics and DL Models

Our analysis revealed substantial biases, both in machine learning models and the CL clinical baseline. Some biases were common: The radar plots (Fig. 1) for

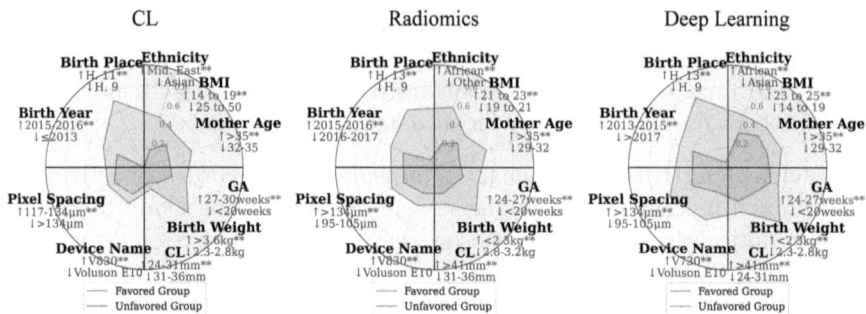

Fig. 1. Which Preterm Prediction Model is Biased, and Where? Shift in sensitivity at 85% specificity between the most favored (blue) and unfavored (red) group for various attributes. CL: cervical length; BMI: body mass index; GA: gestational age. * and ** indicate p < .05 and .001, respectively. (Color figure online)

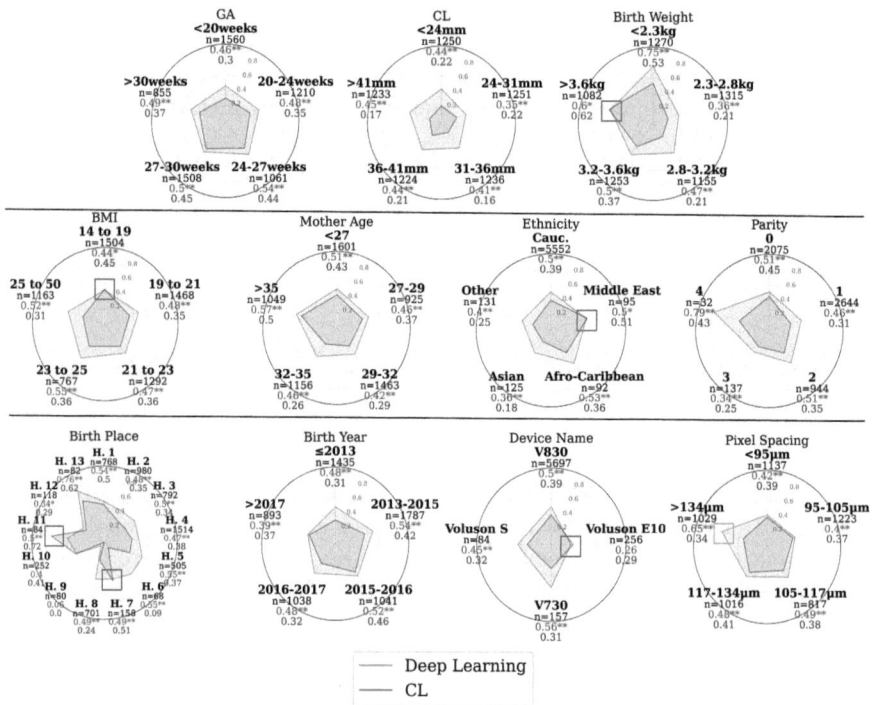

Fig. 2. Where Does Deep Learning Improve Cervical Length Preterm Birth Prediction? Comparison of deep learning and cervical length models on every subgroup in terms of sensitivity at 85% specificity. CL: cervical length; BMI: body mass index; GA: gestational age; Cauc.: Caucasian. * and ** indicate p < .05 and .001.

all three models display similar overall shapes, particularly around birth place and birth weight. The DL model achieved the highest overall performance, its least (red) and most (blue) favored group radars covered the largest areas.

Model-specific biases also emerged. The CL model exhibited stronger performance drops across certain physiological axis such as maternal age, ethnicity, gestational age, and BMI. In contrast, the DL model was more sensitive to imaging-related attributes, such as ultrasound device model and pixel spacing.

3.2 CL and DL Comparison

The **DL model** outperformed the CL model in terms of sensitivity at 85% specificity across nearly all subgroups (Fig. 2). The superiority of DL stood out for cases with early gestational age, long CL measurements (>41 mm), low birth weight, high BMI, middle-aged mothers, and high pixel spacing. **CL performance** varied sharply across physiological strata, confirming its known sensitivity to maternal and fetal profiles. CL slightly outperformed DL in a few subgroups including low BMI (<19), high birth weight (>3.6 kg) and Voluson E10, where DL notably dropped in accuracy compared to other US devices. DL had few Voluson E10 training samples (n = 256).

3.3 Pixel Spacing and DL Performance

The bias previously observed with respect to pixel spacing in the DL model revealed itself as a performance advantage in the subgroup with the highest spacings. In the upper 20% quantile, the DL model achieved an AUC of 0.81

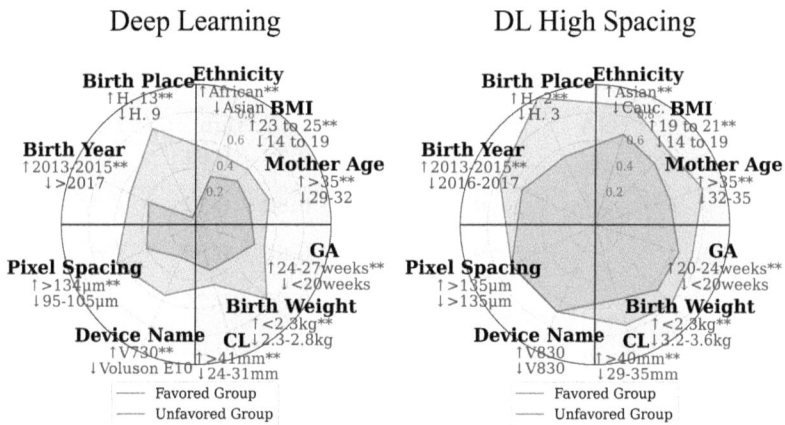

Fig. 3. High Spacing Boosts DL Performance, Mitigates Bias. DL maximum and minimum sensitivity across all pixel spacings (left, n = 7862) or spacing >134 μm (right, n = 1179). CL: cervical length; BMI: body mass index; GA: gestational age. * and ** indicate $p < .05$ and $.001$, respectively.

Table 1. Deep Learning AUC and sensitivity (at 85% specificity) grouped by pixel spacing quantiles.

Spacing Quantile	AUCROC	Sensitivity
0–20%	0.71	0.42
20–40%	0.69**	0.40**
40–60%	0.71**	0.48**
60–80%	0.74**	0.48
80–100%	**0.81****	**0.67****

** Significantly different from previous row ($p < .001$).

and a sensitivity of 0.67 at 85% specificity (Table 1, Fig. 3). This higher pixel spacing was associated with ultrasound images exhibiting increased depth and a broader anatomical context surrounding the cervix (Fig. 4). Notably, model confidence scores increased proportionally with pixel spacing (Fig. 5).

To further investigate this, we identified 172 cases with an available pair of both high and low pixel spacing ultrasounds. There was disagreement between the two associated binary prediction in 55 out of the 172 occurrences, and in 76% of those disagreement cases: the high spacing prediction was correct. Importantly, the observed performance gain extended across all gestational ages, including early screenings at or before 20 weeks of gestation (Table 2), indicating that the observation is not just due to bias in operator training across gestational ages.

In contrast, the CL sensitivity (at 0.85 specificity) went from 0.41 globally to 0.36 on the high spacing subgroup.

Table 2. High spacing effect on AUC and sensitivity grouped by gestational age.

GA	DL	DL with High Spacing**
≤20 weeks	0.70 / 0.46 (n=2132)	**0.79 / 0.61** (n=462)
20–24 weeks	0.72 / 0.48 (n=1842)	**0.82 / 0.72** (n=211)
24–27 weeks	0.75 / 0.54 (n=1508)	**0.85 / 0.69** (n=137)
27–30 weeks	0.77 / 0.50 (n=1284)	**0.86 / 0.76** (n=232)
>30 weeks	0.77 / 0.49 (n=1096)	**0.82 / 0.65** (n=136)

** All differences with DL were significant ($p < .001$).

4 Discussion

Our findings indicate that (1) DL outperforms CL for sPTB prediction on nearly all subgroups, except for an US machine under-represented in its training set; and (2) DL performance improves markedly when the image extends beyond the cervix; including surrounding anatomical context.

To our knowledge, both insights are novel contributions to the field of sPTB prediction. We demonstrate that while imaging domain shifts affect DL models, maternal and fetal physiology can cause performance variability in CL models. The comparative bias profiles we provide (Figs. 1, 2) can support clinical decision-making by guiding obstetricians in determining when to trust an automated risk score versus relying on CL alone.

Some observed biases reflect difficulty differences in prediction – intermediate ranges in CLs, maternal ages, birth weights, or early GA. Birth location bias may also reflect difficulty variations, such as Hospital 9 where preterm cases had higher average CLs than term cases (41 mm vs. 39 mm). But skill differences in US acquisition could also be responsible. Regarding DL, lower performance on under-represented US machines is consistent with known issues of domain shift in medical imaging. This is particularly important: the replacement frequency of US devices – 7 to 9 years – is the shortest among imaging modalities [1].

No other bias assessments have been conducted in sPTB prediction, but our ranking of models (CL < radiomics < DL) aligns with previous literature. This superiority of DL may be attributed to its ability in exploiting imaging patterns beyond cervical length alone, including tissue heterogeneity, canal shape,

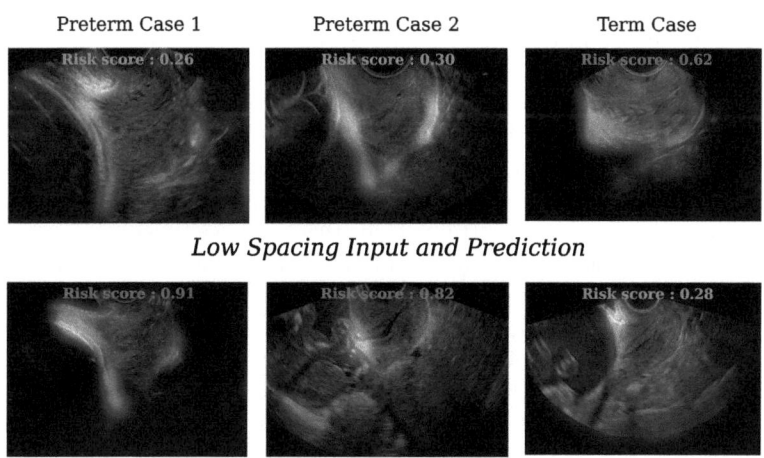

Fig. 4. Changes in Saliency Map and Risk Score Induced by Pixel Spacing Shift. The three columns correspond to three pregnancies and examinations. Each column contains two ultrasounds from the same session: a low spacing (first row), and a high spacing (second row) image, with associated risk score (predicted probability).

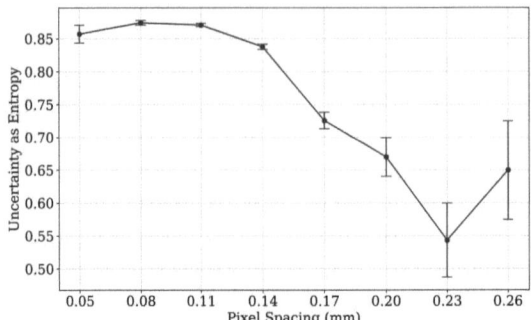

Fig. 5. Increasing Pixel Spacing Increases Model Confidence. Each y-axis value corresponds to the mean entropy (with standard error) for the 0.06 mm wide interval of spacings centered around the x value.

and other morphological or textural cues ignored in standard CL. Our observation that greater pixel spacing improves DL prediction performance extends this logic of looking beyond CL—e.g., by incorporating cervical texture analysis—to enhance predictive accuracy. The observed improvement may stem from the model utilizing broader anatomical context: the shape of the lower uterine segment, the echogenicity of the amniotic fluid, the placental morphology – all of which may relate to different aetiologies of preterm birth.

Limitations of this study include potential dataset selection bias which could e.g. explain the high spacing effect. One might suspect that high spacing US corresponds to easy cases given to unexperienced sonographers. Nonetheless, the large sample size in the high spacing group with comparable maternal and fetal characteristics (n = 1159); the consistency of the performance gain across all ranges of gestational ages (Table 2); the 76% rate of correctness of high spacing over low spacing prediction in 55 disagreement cases; the fact that the CL sensitivity actually got worse on the same subgroup (0.41 to 0.36); all support the robustness of this preliminary finding. Other limitations include low sample sizes in some subgroups (e.g., certain ethnicities or device types). Yet, it is worth noting that these subgroups often contain more cases than the total dataset size in many prior sPTB prediction studies.

By comparing DL and CL performance across real-world subgroups, this work introduces a much-needed level of transparency for automated sPTB prediction to benefit newborns in practice. While this a step forward, further analysis on larger cohorts should explore the behavior of the DL model before considering its clinical adoption, especially with regard to their sensibility to ultrasound devices getting older or replaced. The identified spacing effect is a promising finding which – if confirmed in future studies comparing low versus high spacing within the same patients and prospectively evaluating different scanning protocols – could lead to a simple and low-cost clinical recommendation: Acquire transvaginal US images with extended anatomical context, using DL models for risk prediction. Given the worldwide 13 million annual preterm births, increasing sensitivity from 40% to 60% could improve neonatal outcomes on a global scale.

Acknowledgments. The work was partly funded by the Novo Nordisk foundation through the projects NNF0087102, NNF0062606, and NNF0083562.

Disclosure of Interests. The authors have no competing interests.

References

1. European Society of Radiology: Renewal of radiological equipment. Insights Imaging **5**(5), 543–546 (2014). https://doi.org/10.1007/s13244-014-0345-1
2. Lancet Child Adolesc. Health **6**(1), e4 (2022). https://doi.org/10.1016/s2352-4642(21)00382-5
3. Baños, N., et al.: Quantitative analysis of cervical texture by ultrasound in midpregnancy and association with spontaneous preterm birth. Ultrasound Obstet. Gynecol. **51**(5), 637–643 (2018). https://doi.org/10.1002/uog.17525
4. Burgos-Artizzu, X.P., et al.: Mid-trimester prediction of spontaneous preterm birth with automated cervical quantitative ultrasound texture analysis and cervical length: a prospective study. Sci. Rep. **11**(1), 7469 (2021)
5. Cancino, W., Becerra-Mojica, C.H., Pertuz, S.: Radiomic analysis of transvaginal ultrasound cervical images for prediction of preterm birth. In: Yap, M.H., Kendrick, C., Behera, A., Cootes, T., Zwiggelaar, R. (eds.) MIUA 2024. LNCS, vol. 14860, p. 414–424. Springer, Cham (2024). https://doi.org/10.1007/978-3-031-66958-3_30
6. Coutinho, C.M., et al.: ISUOG practice guidelines: role of ultrasound in the prediction of spontaneous preterm birth. Ultrasound Obstet. Gynecol. **60**(3), 435–456 (2022). https://doi.org/10.1002/uog.26020
7. Drukker, L., Noble, J.A., Papageorghiou, A.T.: Introduction to artificial intelligence in ultrasound imaging in obstetrics and gynecology. Ultrasound Obstet. Gynecol. **56**(4), 498–505 (2020). https://doi.org/10.1002/uog.22122
8. D'Antonio, F., Eltaweel, N., Prasad, S., Flacco, M.E., Manzoli, L., Khalil, A.: Cervical cerclage for prevention of preterm birth and adverse perinatal outcome in twin pregnancies with short cervical length or cervical dilatation: a systematic review and meta-analysis. PLoS Med. **20**(8), e1004266 (2023). https://doi.org/10.1371/journal.pmed.1004266
9. Fiset, S., Martel, A.L., Glanc, P., Barrett, J., Melamed, N.: Prediction of spontaneous preterm birth among twin gestations using machine learning and texture analysis of cervical ultrasound images. Univ. Toronto Med. J. **96**, 6–9 (2019). https://api.semanticscholar.org/CorpusID:201159408
10. Fonseca, E.B., Celik, E., Parra, M., Singh, M., Nicolaides, K.H.: Progesterone and the risk of preterm birth among women with a short cervix. N. Engl. J. Med. **357**(5), 462–469 (2007). https://doi.org/10.1056/nejmoa067815
11. van Griethuysen, J.J., et al.: Computational radiomics system to decode the radiographic phenotype. Can. Res. **77**(21), e104–e107 (2017). https://doi.org/10.1158/0008-5472.CAN-17-0339
12. Iams, J.D., et al.: The length of the cervix and the risk of spontaneous premature delivery. N. Engl. J. Med. **334**(9), 567–573 (1996). https://doi.org/10.1056/nejm199602293340904

13. Mindee: DocTR: Document Text Recognition (2025). https://mindee.github.io/doctr/. Accessed 23 June 2025
14. Myrhaug, H.T., Brurberg, K.G., Hov, L., Markestad, T.: Survival and impairment of extremely premature infants: a meta-analysis. Pediatrics **143**(2) (2019). https://doi.org/10.1542/peds.2018-0933
15. Ohtaka, A., Akazawa, M., Hashimoto, K.: Deep learning algorithm for predicting preterm birth in the case of threatened preterm labor admissions using transvaginal ultrasound. J. Med. Ultrason. **51**(2), 323–330 (2023). https://doi.org/10.1007/s10396-023-01394-9
16. Ohuma, E.O., et al.: National, regional, and global estimates of preterm birth in 2020, with trends from 2010: a systematic analysis. Lancet **402**(10409), 1261–1271 (2023). https://doi.org/10.1016/s0140-6736(23)00878-4
17. Pegios, P., et al.: Leveraging shape and spatial information for spontaneous preterm birth prediction. In: Kainz, B., Noble, A., Schnabel, J., Khanal, B., Müller, J.P., Day, T. (eds.) ASMUS 2023. LNCS, vol. 14337, pp. 57–67. Springer, Cham (2023). https://doi.org/10.1007/978-3-031-44521-7_6
18. Van der Walt, S., et al.: scikit-image: image processing in Python. PeerJ **2**, e453 (2014)
19. Włodarczyk, T., et al.: Spontaneous preterm birth prediction using convolutional neural networks. In: Hu, Y., et al. (eds.) ASMUS/PIPPI -2020. LNCS, vol. 12437, pp. 274–283. Springer, Cham (2020). https://doi.org/10.1007/978-3-030-60334-2_27

Identifying Gender-Specific Visual Bias Signals in Skin Lesion Classification

Heejae Lee[1], Sejung Yang[1], Yuseong Chu[1(✉)], and Byungho Oh[2(✉)]

[1] Department of Precision Medicine, Yonsei University Wonju College of Medicine, Wonju, South Korea
dbtjd9968@yonsei.ac.kr
[2] Department of Dermatology, Severance Hospital, Cutaneous Biology Research Institute, Yonsei University College of Medicine, Seoul, South Korea
obh505@gmail.com

Abstract. Recent advances in AI-based medical diagnosis have demonstrated impressive accuracy. However, concerns remain regarding the fairness of these models across demographic groups. Gender-related biases embedded in skin lesion images may compromise diagnostic equity and lead to systematic disparities in clinical decision making. In this study, we investigate gender-specific visual bias signals in skin lesion classification using the International Skin Imaging Collaboration 2019 dataset. We trained ConvNeXt models on male-only, female-only, and mixed-gender datasets, and evaluated them across all test sets. In addition, we selected 100 samples per signal from our dataset based on visual prominence scores across 10 dermoscopic features known to influence skin lesion appearance. Our results reveal systematic disparities in model performance across gender groups. For example, the Blue signal led to a sharp performance drop when the female model was evaluated on male data, while the male model performed substantially better on the same subset. This contrast highlights how certain visual signals can hinder cross-gender generalization. These findings suggest that certain visual features affect model reliability depending on patient gender, raising concerns for fairness in real-world clinical deployment. Our work provides empirical evidence and diagnostic insights that can support the development of bias-aware dermatological AI systems.

Keywords: Gender Bias · Model Fairness in Medical AI · Skin Lesion Classification

1 Introduction

Skin cancer constitutes a major portion of cancer diagnoses globally, with melanoma and non-melanoma types posing serious health threats [8]. Early and accurate diagnosis is necessary for effective treatment and survival. Numerous AI-based studies have demonstrated high diagnostic accuracy in detecting skin cancers and other dermatological diseases using dermoscopic images [3,6,18]. Despite recent successes, deep learning models for skin lesion diagnosis may

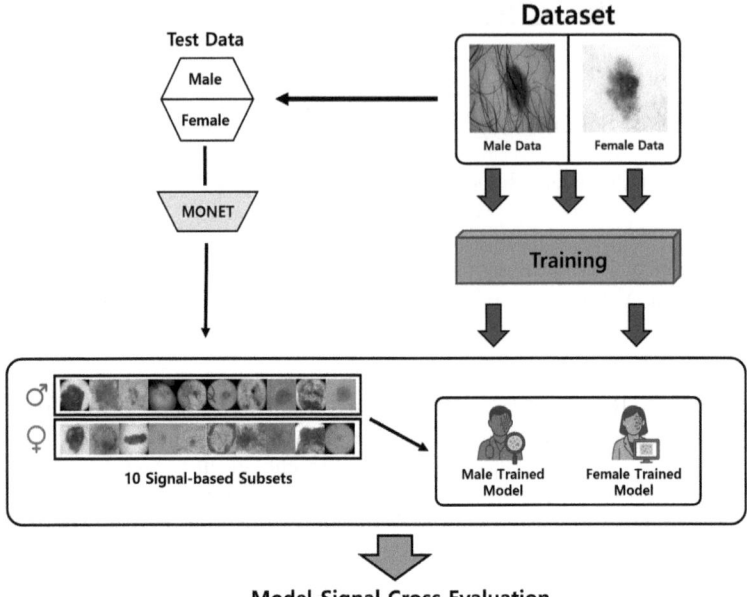

Fig. 1. Overview of the experimental workflow for gender signal evaluation in skin lesion classification.

unintentionally learn biased features that degrade performance or cause model instability, even as they capture clinically relevant patterns.

While much of the existing research has focused on algorithmic biases related to skin tones, emerging studies suggest that gender-related factors may also affect model performance–not just in dermatology but across a wide range of medical imaging tasks. In radiology and MRI-based diagnostics, studies have shown that AI models can encode and respond differently to gender. For instance, classifiers trained on gender-imbalanced chest X-ray datasets yielded higher performance on male patients but underperformed on female cases [13]. Similarly, in Alzheimer's disease diagnosis using brain MRI, sex-related imaging features were shown to influence model performance and generalizability, even when sex is not explicitly annotated [16].

Recent work in dermatology further highlights the risk of gender-specific bias [7]. Banerjee et al. demonstrated that deep learning models trained on skin lesion images could encode gender in latent feature representations, even when gender was not explicitly labeled. Their study found that visual characteristics such as hair, skin tone, pigmentation, and texture may act as inadvertent proxies for gender, resulting in gender-dependent model behavior and potentially reducing diagnostic fairness [2]. Raumanns et al. demonstrated that the distribution of patient sex in skin lesion datasets strongly affects model fairness, and revealed systematic gender bias in model performance across different training strategies and dataset configurations [17].

Motivated by prior findings, our study aims to examine whether gender-specific training of skin lesion classification models produce gender-dependent performance differences. Using the International Skin Imaging Collaboration (ISIC) 2019 dataset [19], we constructed three training subsets: male, female, and fair-gender. We then trained separate models on each subset and conducted cross-evaluation to assess gender-specific performance differences. Furthermore, to examine the impact of potential bias inducing signals, we extracted ten distinct visual signals from male and female test sets and evaluated them using models trained on both genders. An overview of the experimental design is illustrated in Fig. 1. This design allows us to identify which signals most strongly contribute to gender related performance disparities.

2 Methods

2.1 Dataset

The dataset used in this study is the ISIC 2019 Challenge Dataset, a large-scale publicly available collection of dermoscopic images for automated skin lesion diagnosis. It consists of 25,331 training images and 8,238 test images, categorized into nine diagnostic classes of skin lesions. The ISIC 2019 Challenge Dataset is part of the International Skin Imaging Collaboration (ISIC) archive and incorporates multiple sources, including the HAM10000 dataset [19]. We selected the ISIC 2019 Challenge Dataset for this study due to its large size, clinical diversity, and public availability, which make it a benchmark dataset for skin lesion classification. Importantly, the dataset includes metadata such as patient sex, which allows for gender-based sub-group analysis.

In this study, we focused on three clinically relevant lesion types from the ISIC 2019 dataset: melanoma (MEL), melanocytic nevus (NV), and basal cell carcinoma (BCC). These categories were selected based on their clinical significance, relatively large sample sizes, and visual characteristics, which are crucial for consistent model training and subgroup fairness evaluation. Moreover, these three classes are among the most encountered and diagnostically critical lesions in real-world dermatological practice, making them an appropriate target for analyzing model bias and generalizability [11].

We constructed two gender-specific datasets by selecting only the samples with available gender labels from the original dataset. The male dataset comprises 10,686 training images and 2,314 test images, and the female dataset consists of 9,676 training images and 2,271 test images. Although the male and female datasets are reasonably balanced, a slight discrepancy in sample sizes remains, potentially reflecting real-world collection bias. Prior to validation splitting, the male training dataset consisted of 2,461 MEL, 6,225 NV, and 2,000 BCC images, while the female training dataset comprised 1,980 MEL, 6,379 NV, and 1,317 BCC images. This distribution highlights a significant class imbalance, with NV being substantially overrepresented in both gender subsets.

2.2 Model

For experiment, we implemented ConvNeXt [14], a modern convolutional neural network architecture that integrates design principles from Vision Transformers into a pure CNN framework. ConvNeXt offers improved representation capacity and optimization stability over traditional architectures such as ResNet [9,20], making it well-suited for medical image classification tasks like dermoscopic skin lesion analysis. Its strong performance on natural image benchmarks and robust feature extraction capabilities motivated its selection for our gender-sensitive fairness experiments [14].

We conducted experiments using three models trained with ConvNeXt: a female model trained exclusively on female data, a male model trained exclusively on male data, and a fair model trained on both male and female data. Each model was evaluated on three test sets: female, male, and fair gender, resulting in a total of nine evaluation scenarios. For all evaluations, we applied the conventional threshold of 0.5 to the predicted probabilities for class assignment. To train the ConvNeXt-Base model, we employed a class-weighted cross-entropy loss to address the class imbalance among MEL, NV, and BCC cases. Specifically, the class weights were computed as the inverse square root of class frequencies, a strategy designed to reduce the dominance of the overrepresented NV class while maintaining stability for minority classes [10]. For data augmentation, we applied minimal transformations to preserve the visual features and demographic signals of the dermoscopic images [2]. A random rotation of 90 or 180 degrees was applied with a probability of 0.6 to a subset of training images, while no other synthetic augmentation was used. All images were normalized using ImageNet [5] standard mean and standard deviation values. For validation and test sets, only normalization was applied to ensure consistency in evaluation.

Model training was conducted using the Adam optimizer with a fixed learning rate of 1×10^{-5}. The batch size was set to 16, and training was carried out for up to 100 epochs. To improve training stability, gradient clipping was applied with a maximum norm of 5.0. Early stopping was employed when the validation loss did not improve for five consecutive epochs. While random seeds were not fixed in this study, all experiments were conducted using consistent hardware and software environments to minimize uncontrolled variability.

2.3 MONET-Based Visual Attribute Group Extraction

To facilitate detailed analysis of model performance across clinically relevant image characteristics, we utilized MONET (Medical cONcept rETriever), a vision-language foundation model capable of identifying dermatological attributes from medical images [12]. MONET projects both visual inputs and textual attribute prompts into a unified embedding space via contrastive learning, enabling quantification of attribute relevance through similarity scores.

Moreover, MONET demonstrated robust zero-shot concept transfer: in an evaluation on ISIC dermoscopic images, it achieved clinically meaningful concept extraction and classification performance via concept differential analysis and concept bottleneck model evaluation , and on a Fitzpatrick17k dataset encompassing diverse skin tones, it outperformed non-medical CLIP in zero-shot concept extraction accuracy while maintaining consistency across skin types.

Using this framework, we organized the test dataset into visually coherent subsets by querying MONET with a curated list of clinically meaningful concepts. For each concept we retrieved the top 100 images exhibiting the highest similarity scores, thus ensuring strong visual presence of the specified attribute without requiring manual annotation.

3 Experiments and Results

3.1 Experimental Setup

For each gender-specific dataset, the corresponding test set was kept unchanged. From the training set, approximately 15% of the samples were held out as a validation set to monitor performance and prevent overfitting. As a result, the final proportions of training, validation, and test data were approximately 69%, 12%, and 19%, respectively.

To retain the intrinsic visual characteristics of the lesion images, only minimal preprocessing was applied. All images were uniformly resized to 224×224 pixels. No synthetic data augmentation techniques were used for validation or test sets; these subsets were only normalized using ImageNet mean and standard deviation. For training, minimal augmentation was applied (e.g., random rotations), but techniques such as color jittering or zoom were deliberately avoided. This decision was motivated by recent fairness studies [2,13], which highlight that such augmentations can unintentionally change or obscure latent demographic signals embedded in medical images. In our study, preserving these subtle features was critical for accurately evaluating gender-related performance disparities.

3.2 Baseline Fairness Evaluation

We analyzed the performance of three models across different gender-specific test sets to investigate how training data composition affects generalization and class-wise accuracy. A summary of cross-evaluation accuracy and AUROC scores across all models and test sets is presented in Table 1.

The female model exhibited better generalization to the male test set than to its own female test set in MEL classification. However, the lowest performance was observed on female MEL cases ($71.5 \pm 4.4\%$), indicating potential intra-gender overfitting or gender-specific underrepresentation [13]. This result suggests that MEL lesions in female skin may have not been adequately learned or were underrepresented during training [2,18]. In contrast, NV accuracy remained consistently high across all test sets, suggesting robustness and stability of this class across genders.

The male model performed best when evaluated on its own male test set (Accuracy: $82.7\pm0.9\%$, MEL: $74.5\pm3.5\%$), highlighting the advantage of gender-aligned training. However, when applied to the female test set, its overall accuracy dropped by approximately 3%, and NV performance decreased slightly (by

Table 1. AUROC and Accuracy results from cross-evaluation of gender-specific and fair models on male, female, and mixed test sets.

Model	Test Set	AUROC	Accuracy (%)	ACC_MEL (%)	ACC_NV (%)	ACC_BCC (%)
Female Model	Female	0.932 ± 0.005	81.1 ± 0.9	71.5 ± 4.4	84.7 ± 1.9	86.0 ± 3.4
	Male	0.938 ± 0.002	81.5 ± 1.1	75.3 ± 1.6	85.1 ± 1.9	81.2 ± 3.7
	Fair	0.934 ± 0.003	81.3 ± 0.6	73.3 ± 2.7	84.9 ± 1.9	83.0 ± 3.4
Male Model	Female	0.925 ± 0.002	80.0 ± 0.4	68.6 ± 3.4	84.2 ± 2.1	86.6 ± 1.2
	Male	0.943 ± 0.002	82.7 ± 0.9	74.5 ± 3.5	84.3 ± 2.6	88.7 ± 2.3
	Fair	0.935 ± 0.001	81.4 ± 0.6	71.5 ± 3.4	84.2 ± 2.3	87.9 ± 1.8
Fair Model	Female	0.936 ± 0.002	82.5 ± 0.6	71.6 ± 2.2	86.5 ± 1.5	88.0 ± 2.6
	Male	0.950 ± 0.003	84.1 ± 0.5	75.9 ± 2.0	87.9 ± 1.0	85.8 ± 3.9
	Fair	0.944 ± 0.003	83.3 ± 0.4	73.7 ± 2.1	87.2 ± 1.2	86.7 ± 3.3

1%), suggesting a marginal effect of gender-aligned training. This performance gap suggests that the male model is more specialized to male-specific visual representations and is less generalizable to female lesion patterns [14]. The NV classification accuracy of the male model was around 84%, remaining slightly below that of the other models. This may indicate that the interpretation of NV features is more gender-sensitive than previously expected.

The fair model exhibited the most stable and generalized performance across all test sets. It achieved the highest overall accuracy and AUROC, particularly on the male test set (Accuracy: 84.1 ± 0.5%, AUROC: 0.950 ± 0.003). Notably, both NV and BCC classification accuracies remained consistently high across all test sets, while MEL classification showed relatively lower performance compared to these lesion types. These results suggest that the fair model offers the most balanced generalization across genders and lesion categories, despite persistent challenges in MEL classification.

3.3 Signal-Based Evaluation

To investigate model sensitivity to visual features associated with gender bias, we selected ten visual signals from the test dataset. These include color features (Black, Blue, Pigmentation, Redness, Yellow) [15], structural features (Border Irregularity, Ulceration) [1], and artifacts or non-lesion patterns (Hair, Pen & Marker, Dermoscope Dark Border) [4]. We constructed ten signal-specific subsets for each gender (20 subsets in total). Using these subsets, we conducted cross-evaluations to examine the effect of each visual signal on classification performance across genders.

In these signal-specific test subsets, the distribution of lesion types varied: in the female subsets, the distribution of lesion types was 32.4% MEL, 16.4% BCC and 51.2% NV. In the male subsets, the proportions were 25.0% MEL, 33.1% BCC and 41.9% NV. Despite the potential correlation between certain signals and lesion characteristics, we chose not to enforce strict lesion-type balance in the signal-specific subsets, prioritizing the evaluation of visual signals across genders.

Overall Accuracy on Signal Subsets. Across all signal-based subsets, both models demonstrated comparable overall performance, with average accuracies of 82.4% (female model on female data), 84.3% (female model on male data), 83.8% (male model on male data), and 83.1% (male model on female data). These results suggest that both models are capable of generalizing across gender-specific signal distributions, although some variation remains depending on the model–dataset pairing.

Lesion-Wise Performance. Performance differed substantially across lesion types. For MEL, both models yielded relatively low average accuracies, with the male model slightly outperforming the female model (67.0% vs. 64.0%). For BCC, the male model showed superior performance (86.6% vs. 73.7%), particularly on signals such as Redness, Ulceration, and Yellow. In contrast, for NV, the female model outperformed the male model (87.3% vs. 79.5%), especially on signals including Hair, Pen & Marker, and Border Irregularity. These results highlight lesion-specific variation in model performance, indicating differential vulnerability to signal-induced bias.

Cross-Gender Evaluation Summary. When evaluated on the same-gender signal subsets, each model achieved its highest performance: 84.8% for the male model on male data and 82.3% for the female model on female data. Notably, the female model achieved comparable performance on male signal subsets (83.9%), suggesting a relatively higher degree of generalization. Meanwhile, the male model exhibited a small performance drop (-1.7%) when evaluated on female signal data. While the overall difference across configurations remained modest (above 82% in all cases), these results show that each model generalizes differently depending on gender and highlight the need for closer analysis of signal level behavior.

Signal-Specific Performance Gaps. Performance on certain signals showed variation depending on the model and the gender of the dataset. Figure 2 illustrates the accuracy variations across 10 signal-specific subsets using both male and female models tested on gender-specific data. For example, in the Blue subset, the female model's accuracy dropped from 87% (on female data) to 78% (on male data), while the male model improved from 80% to 88%. This indicates that Blue may encode gender-specific visual features, leading to divergent model responses depending on the training set. Similarly, Redness exhibited an unusual performance pattern: the female model performed better on male data (94%) than on its own female data (82%), while the male model showed stable performance (92%) across genders. Ulceration also demonstrated a surprising pattern: the male model achieved 92% on female data, outperforming its accuracy on male data (87%), whereas the female model remained stable but lower (8182%). These anomalies point to latent gender-specific representations that can emerge even in models not explicitly trained to capture them.

Overall, these findings underscore that visual signals in dermoscopic images can affect model performance based on gender, and that some signals (e.g., Blue, Redness, Ulceration) may act as bias-inducing factors that reduce the fairness and generalizability of classification models.

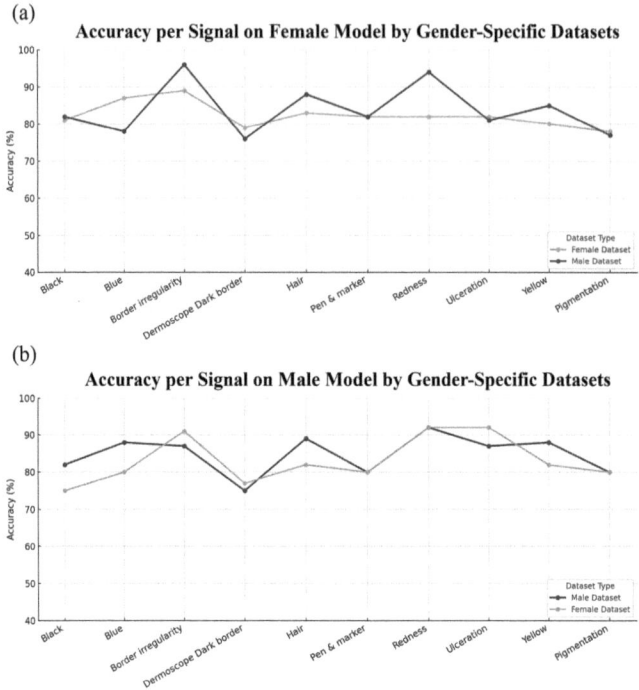

Fig. 2. Signal-wise accuracy comparison of Female and Male models on gender-specific datasets. (a): Female model tested on male and female data. (b): Male model tested on male and female data. Results highlight how specific signals lead to divergent accuracy patterns.

4 Conclusion

This study investigated how gender-specific models respond to various visual signals in the context of skin lesion classification, aiming to uncover potential gender-induced biases embedded in model behavior. Using ConvNeXt-based models trained separately on male and female datasets, we conducted a signal level cross-evaluation across ten commonly observed dermoscopic signals. Our findings identified several signals that may contribute to gender-specific prediction bias. Redness, Ulceration, and Blue led to accuracy discrepancies exceeding 10% across modeldata combinations, indicating increased responsiveness to gender-specific visual patterns. Additionally, performance degradation in cross-model evaluations–such as the Male Model's drop on Blue signals in female data–highlights the risk of biased feature encoding. These signals should be carefully considered when designing fair and generalizable models for dermatological diagnostics. This signal-based evaluation highlights how visual features can affect model behavior in complex, gender-dependent ways. Recognizing and mitigating such gender-specific signals will be essential for developing fair and trustworthy medical AI systems.

Acknowledgment. This research was supported by the National Research Foundation of Korea (NRF) grant funded by the Korea government (MSIT) (2022R1A2C2091160) and by the Bio & Medical Technology Development Program of the National Research Foundation (NRF), also funded by the Korean government (MSIT) (RS-2024-00440802).

Disclosure of Interests.. This manuscript reflects only the authors' views and opinions; neither the European Union nor the European Commission can be considered responsible for them. The authors have no competing interests to declare that are relevant to the content of this article.

References

1. Ali, A.R., Li, J., Yang, G., O'Shea, S.J.: A machine learning approach to automatic detection of irregularity in skin lesion border using dermoscopic images. PeerJ Comput. Sci. **6**, e268 (2020)
2. Banerjee, I., et al.: Reading race: Ai recognises patient's racial identity in medical images. arXiv preprint arXiv:2107.10356 (2021)
3. Brinker, T.J., et al.: Deep neural networks are superior to dermatologists in melanoma image classification. Eur. J. Cancer **119**, 11–17 (2019)
4. Delibasis, K., Moutselos, K., Vorgiazidou, E., Maglogiannis, I.: Automated hair removal in dermoscopy images using shallow and deep learning neural architectures. Comput. Methods Programs Biomed. Update **4**, 100109 (2023)
5. Deng, J., Dong, W., Socher, R., Li, L.J., Li, K., Fei-Fei, L.: Imagenet: a large-scale hierarchical image database. In: 2009 IEEE Conference on Computer vision and Pattern Recognition, pp. 248–255. IEEE (2009)
6. Esteva, A., et al.: Dermatologist-level classification of skin cancer with deep neural networks. Nature **542**(7639), 115–118 (2017)
7. Gadgil, S., DeGrave, A.J., Daneshjou, R., Lee, S.I.: Discovering mechanisms underlying medical ai prediction of protected attributes. medRxiv, pp. 2024–04 (2024)
8. Garbe, C., et al.: Diagnosis and treatment of melanoma. european consensus-based interdisciplinary guideline–update 2016. Euro. J. Cancer **63**, 201–217 (2016)
9. He, K., Zhang, X., Ren, S., Sun, J.: Deep residual learning for image recognition. In: Proceedings of the IEEE Conference on Computer Vision and Pattern Recognition, pp. 770–778 (2016)
10. Johnson, J.M., Khoshgoftaar, T.M.: Survey on deep learning with class imbalance. J. Big Data **6**(1), 1–54 (2019)
11. Kawahara, J., BenTaieb, A., Hamarneh, G.: Deep features to classify skin lesions. In: 2016 IEEE 13th International Symposium on Biomedical Imaging (ISBI), pp. 1397–1400. IEEE (2016)
12. Kim, C., et al.: Transparent medical image ai via an image-text foundation model grounded in medical literature. Nat. Med. **30**(4), 1154–1165 (2024)
13. Larrazabal, A.J., Nieto, N., Peterson, V., Milone, D.H., Ferrante, E.: Gender imbalance in medical imaging datasets produces biased classifiers for computer-aided diagnosis. Proc. Natl. Acad. Sci. **117**(23), 12592–12594 (2020)

14. Liu, Z., Mao, H., Wu, C.Y., Feichtenhofer, C., Darrell, T., Xie, S.: A convnet for the 2020s. In: Proceedings of the IEEE/CVF Conference on Computer Vision and Pattern Recognition, pp. 11976–11986 (2022)
15. Massi, D., De Giorgi, V., Carli, P., Santucci, M.: Diagnostic significance of the blue hue in dermoscopy of melanocytic lesions: a dermoscopic-pathologic study. Am. J. Dermatopathol. **23**(5), 463–469 (2001)
16. Petersen, E., et al.: Feature robustness and sex differences in medical imaging: a case study in mri-based alzheimer's disease detection. In: International Conference on Medical Image Computing and Computer-Assisted Intervention. pp. 88–98. Springer (2022). https://doi.org/10.1007/978-3-031-16431-6_9
17. Raumanns, R., Schouten, G., Pluim, J.P., Cheplygina, V.: Dataset distribution impacts model fairness: Single vs. multi-task learning. In: MICCAI Workshop on Fairness of AI in Medical Imaging, pp. 14–23. Springer (2024). https://doi.org/10.1007/978-3-031-72787-0_2
18. Tschandl, P., et al.: Human-computer collaboration for skin cancer recognition. Nat. Med. **26**(8), 1229–1234 (2020)
19. Tschandl, P., Rosendahl, C., Kittler, H.: The ham10000 dataset, a large collection of multi-source dermatoscopic images of common pigmented skin lesions. Scientific Data **5**(1), 1–9 (2018)
20. Wu, Y., Chen, B., Zeng, A., Pan, D., Wang, R., Zhao, S.: Skin cancer classification with deep learning: a systematic review. Front. Oncol. **12**, 893972 (2022)

Fairness-Aware Data Augmentation for Cardiac MRI Using Text-Conditioned Diffusion Models

Grzegorz Skorupko[1](✉), Richard Osuala[1,2,3], Zuzanna Szafranowska[1], Kaisar Kushibar[1], Vien Ngoc Dang[1], Nay Aung[4,5], Steffen E. Petersen[4,5], Karim Lekadir[1,6], and Polyxeni Gkontra[1]

[1] Barcelona Artificial Intelligence in Medicine Lab (BCN-AIM), Departament de Matemàtiques i Informàtica, Universitat de Barcelona, Barcelona, Spain
grzegorz.skorupko@ub.edu
[2] Helmholtz Center Munich, Munich, Germany
[3] Technical University of Munich, Munich, Germany
[4] William Harvey Research Institute, NIHR Barts Biomedical Research Centre, Queen Mary University London, Charterhouse Square, London, UK
[5] Barts Heart Centre, St Bartholomew's Hospital, Barts Health NHS Trust, West Smithfield, London, UK
[6] Institució Catalana de Recerca i Estudis Avançats (ICREA), Barcelona, Spain

Abstract. While deep learning holds great promise for disease diagnosis and prognosis in cardiac magnetic resonance imaging, its progress is often constrained by highly imbalanced and biased training datasets. To address this issue, we propose a method to alleviate imbalances inherent in datasets through the generation of synthetic data based on sensitive attributes such as sex, age, body mass index (BMI), and health condition. We adopt ControlNet based on a denoising diffusion probabilistic model to condition on text assembled from patient metadata and cardiac geometry derived from segmentation masks. We assess our method using a large-cohort study from the UK Biobank by evaluating the realism of the generated images using established quantitative metrics. Furthermore, we conduct a downstream classification task aimed at debiasing a classifier by rectifying imbalances within underrepresented groups through synthetically generated samples. Our experiments demonstrate the effectiveness of the proposed approach in mitigating dataset imbalances, such as the scarcity of diagnosed female patients or individuals with normal BMI level suffering from heart failure. This work represents a major step towards the adoption of synthetic data for the development of fair and generalizable models for medical classification tasks. Notably, we conduct all our experiments using a single, consumer-level GPU to highlight the feasibility of our approach within resource-constrained environments. Our code is available at https://github.com/faildeny/debiasing-cardiac-mri.

Keywords: Deep Learning · Generative Models · Bias Mitigation · Cardiac Imaging

1 Introduction

Cardiovascular diseases remain the main cause of mortality worldwide, accounting for approximately one third of annual deaths globally [5]. Cardiovascular magnetic resonance (CMR) is currently the gold standard in evaluating the structure and function of the heart. However, its acquisition is expensive and the annotation process of multi-slice cine sequences requires a significant amount of time. Consequently, the amount of available training data is limited, hindering the adoption of deep learning based algorithms. Despite the efforts to automate CMR dataset collection, annotation and analysis, end-to-end models are still not common. Such solutions are more affected by the inherent biases in the training data especially when the data is scarce. For example, Puyol et al. [15] showed discrepancies in the performance of CMR segmentation models for subgroups based on sex and race. This finding was primarily attributed to the pronounced imbalance in the training dataset, which consisted mostly of individuals of white race. Such biases can significantly influence the decision-making process of classification models and were widely studied and addressed in various medical domains [9,10,14,20,23].

Advancements in generative deep learning models opened paths to previously unexplored approaches in tackling this crucial challenge in machine learning, namely, algorithmic bias. Some studies have proposed bias mitigation methods through different sampling strategies or modifications to model architecture and training procedures [21,24]. Nonetheless, in the medical domain, the adoption of generative models to mitigate biases through the use of synthetic data has received relatively little attention. Recent works based on GANs and Diffusion models focusing on dermatology, chest X-ray and histopathology domains, are among the very few examples in this direction [7,11]. Ktena et al. [7] proposed models conditioned on both diagnostic and sensitive attributes, such as sex, age, or skin tone, allowed to augment the unbalanced training dataset and successfully reduce the biases in classification tasks. However, to the best of our knowledge, none of the previous works focused on magnetic resonance imaging (MRI) or cardiovascular domain, nor did they allow for conditioning image generation on shape information from segmentation masks or textual prompts.

To address this gap, we propose an open-source pipeline involving training of a resource-intensive stable diffusion model [16] within a limited computa-

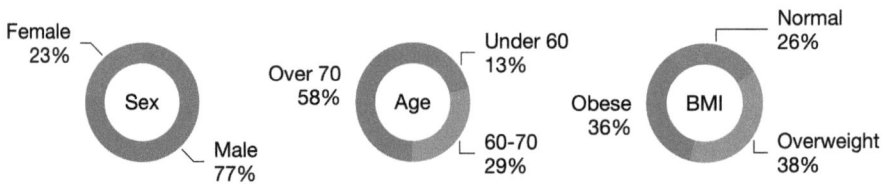

Fig. 1. Demographic statistics of patients diagnosed with heart failure from the UK Biobank imaging study.

tional environment. More precisely, we implement a latent diffusion model with combined text and image inputs to generate spatially consistent CMR frames to mitigate biases introduced by unbalanced training data in CMR-based deep learning models for disease diagnosis. This approach facilitates the generation of CMR data for underrepresented patient subgroups, considering factors such as sex, age, BMI, and heart conditions, alongside spatial-temporal features defined through segmentation masks from multiple cardiac phases. We evaluate the quality of the generated images using the domain-specific, recently introduced [12] and validated [6] Fréchet Radiomics Distance (FRD) score. Furthermore, we assess the impact of the attributed-conditioned synthetic images in heart failure classification model training, demonstrating enhanced model fairness and performance across diverse patient subgroups. Overall, the proposed approach serves as a general-purpose targeted augmentation method, as we illustrate its applicability in resource-limited environments. Our key contributions are:

1. A promising data augmentation method for improving fairness through an Attribute-Conditioned Latent Diffusion Model.
2. The first application of a diffusion model to explicitly address fairness in cardiac MRI, and the first to condition on BMI—an important but underexplored factor in this modality.
3. We experimentally demonstrate that the proposed method simultaneously improves both fairness and classification performance across subgroups, highlighting its potential for clinical adoption.

2 Methodology

2.1 Dataset

For this study, we use the UK Biobank (UKBB) [18], a large-scale resource with data from over 500,000 participants recruited between 2006 and 2010, that includes demographics, electronic health records (EHRs), biomarkers, and genomics. We focus on a subset of patients who participated in the imaging study and underwent CMR scans. In total, our dataset consists of 25480 multi-slice, short-axis cine CMRs with annotations for end-diastole (ED) and end-systole (ES) frames. The annotation masks label key cardiac structures: left and right ventricles and myocardium. Based on International Classification of Diseases (ICD-10) codes from in-hospital patient data, we identified a subset of 270 patients diagnosed with heart failure at the time of the CMR acquisition. Figure 1 provides the distribution of characteristics of the participants included in the study. In our analysis, we divided patients into groups by age: below 60, 60–70 and over 70 years old, by BMI: below 25 (underweight and normal), 25–30 (overweight) and over 30 (obese), and by sex.

Data Pre-processing. Due to the multidimensional nature of CMR samples (4D), we conduct several data preprocessing steps to adapt to the image format most commonly used in state-of-the-art classification models, i.e. 2D, 3-channel images, to be generated by the Stable diffusion model with ControlNet. We extract the central slice from each volume and stack cine frames from ED and ES phases as color channels, creating a 2D RGB image. To keep the advantage of multidimensional data, we extract three central slices per patient and include cine frames before and after ED and ES, increasing training images to a total of 229,320. We do not apply this augmentation to the validation or test sets, where we solely use one central slice with ED and ES frames.

2.2 Conditioned Image Generation

An overview of the proposed pipeline for generating synthetic CMR images based on textual information and cardiac masks is provided in Fig. 2. We use ControlNet [25], which enhances Stable Diffusion [16] by enabling fine-tuning with text and image inputs. The approach duplicates the pretrained model, adding spatial input only to the cloned branch, which connects to the original architecture via zero convolution layers to reduce noise and preserve the trainable copy's backbone. The original model's weights remain locked to retain generative capabilities, allowing adaptation to new imaging domains without costly retraining. To encode clinical attributes such as BMI, we construct prompts using group names (e.g., "obese") instead of continuous values, as direct inclusion in the prompt did not prove effective and using numerical inputs outside the prompt would require architectural changes.

Diffusion Model Training. We conduct all experiments on a single Nvidia 3080Ti GPU with 16.GB of memory. To train the diffusion model, we adopt the implementation by [25]. To fully leverage the advantages of the pretrained model, we upscale the samples to 512×512 pixels to match the final pretraining resolution of a Stable diffusion 2.1-base model [16]. In the training setup, we use the pretrained image AutoEncoder network and the OpenClip [4] text encoder

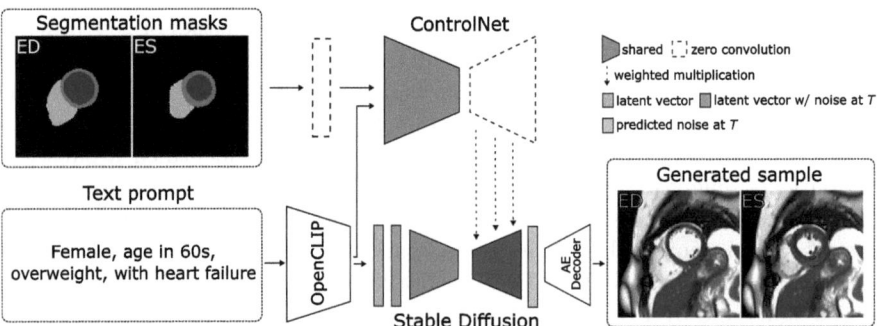

Fig. 2. Overview of the proposed pipeline for generating synthetic CMR data conditioned on textual information and cardiac geometry derived from segmentation masks.

pretrained on the LAION-5B [17] dataset. During the training phase, we exclusively fine-tune the ControlNet branch of the model. In this setup, it is possible to train the model with batch size of 1 with 2 gradient accumulations. We train the model with a learning rate of 1e-5 for 5 epochs, which takes approximately 3 d in our setup. All the code is based on PyTorch framework [13].

Debiased Dataset Generation. To address biases resulting from underrepresentation of certain groups, we use weighted random sampling on our initial dataset. Patients are grouped based on sex, age, BMI and diagnosis, which creates 36 groups in total. For example, female, overweight patients younger than 60 years that are healthy belong to the same subset. Based on each group's population we calculate the sampling weights that are inversely proportional to their size. We subsequently generate synthetic images based on existing prompts and masks for underrepresented groups. This way, we ensure that imaging inputs are coherent with patient's characteristics and do not contribute to additional noise.

2.3 Downstream Classification Model Training

For the downstream classification task, due to the relatively small dataset size, we use a well established ResNet-18 model [2] with weights pretrained on the ImageNet dataset [1]. All training samples are scaled to the native pretraining resolution of 224×224 pixels. Models are trained for 10 epochs with a batch size of 64, starting at a 1e-4 learning rate, reduced by 2 on plateau for 3 epochs. Standard augmentations like random flipping and Gaussian noise are applied. We save model weights after each epoch, selecting the best checkpoint based on balanced accuracy. The dataset is split into 20% test data, with 20% of the training set reserved for validation. During training, we explore different sampling methods, including sample weighting (SW), which adjusts weights based solely on label, and stratified sample weighting (SSW), which considers the joint distribution of subject label and sensitive subgroup. In SSW, the final group sizes vary substantially, with the smallest subgroup containing only 20 images. For each training we use a full set of real images and a proportion of synthetic images depending on the scenario.

2.4 Evaluation Metrics

Synthetic Data Evaluation. To evaluate synthetic medical image quality, we use the radiology domain-specific FRD, thereby avoiding the limitations of alternatives such as the Fréchet Inception Distance (FID) [3], which, pretrained on natural images, often lacks robustness in medical imaging [6,22]. In contrast, FRD measures distances between distributions of radiomics features, which are a proven method for characterizing medical images [6,8,12,19]. To assess the model's ability to condition images on sensitive attributes, we compute FRD within subpopulations (e.g., only females) and between groups (e.g., females vs males). This allows us to evaluate how well real image feature distributions are preserved in data generated by our model.

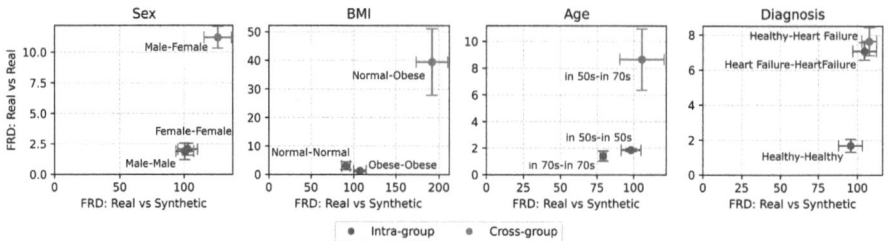

Fig. 3. FRD scores within original dataset (Real vs Real) and with synthetic data (Real vs Synthetic) calculated for attribute subpopulations.

Classification Task. To evaluate classifier performance on heart failure diagnosis, we use AUROC and Balanced Accuracy (BACC), the latter addressing class imbalance due to disease prevalence of ~1%. Metrics are reported globally, per subgroup, and as an average between groups to ensure equal importance across populations, providing a fairer assessment of model performance.

For fairness evaluation, we use the Equal Opportunity Difference (EOD). EOD measures the disparity in true positive rates (TPR) across demographic groups, ensuring that the model performs equally well for all groups in terms of correctly identifying positive outcomes. The formula for EOD is given by:

$$\text{EOD} = \min_{x \in \Omega_X} \text{TPR}_x - \max_{x \in \Omega_X} \text{TPR}_x, \tag{1}$$

where TPR_x represents the true positive rate for group x, and Ω_X denotes the set of all groups under consideration.

3 Results

3.1 Synthetic Data Evaluation

FRD Scores Comparison Between Subpopulations. Figure 3 shows FRD values for CMR images across subgroups categorized by sex, age, BMI, and health condition. The vertical axis (Real vs. Real) captures visual differences in real data, while the horizontal axis (Real vs. Synthetic) evaluates how well these differences are preserved in synthetic images. Intra-group comparisons (Female-Female) yield lower FRD scores, while cross-group (Male-Female) show higher values, indicating expected dissimilarities. Synthetic images have higher FRD scores but follow a similar trend. A comparable pattern appears for BMI, where synthetic images of obese patients closely resemble real high-BMI subjects. For age, real datasets show notable radiomics feature differences, which are less distinct in synthetic images, especially for younger patients. Finally, differences between heart failure and healthy individuals are subtler than for other attributes in both real and synthetic data, highlighting the difficulty of the diagnosis task.

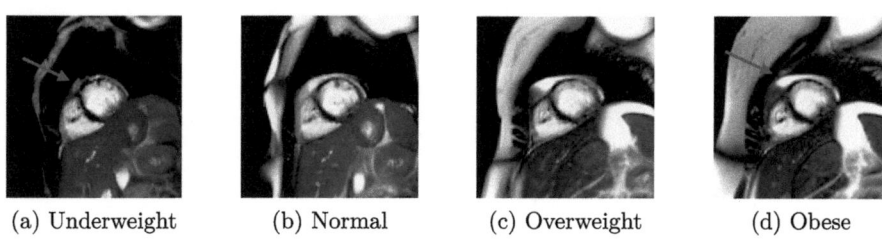

(a) Underweight (b) Normal (c) Overweight (d) Obese

Fig. 4. Effect of altering sensitive features on the generated images using the prompt: "Female, age in 60s, {*BMI category*}". In this example, the sensitive attribute BMI was modified between generation runs to observe its effect on the generated CMR scans.

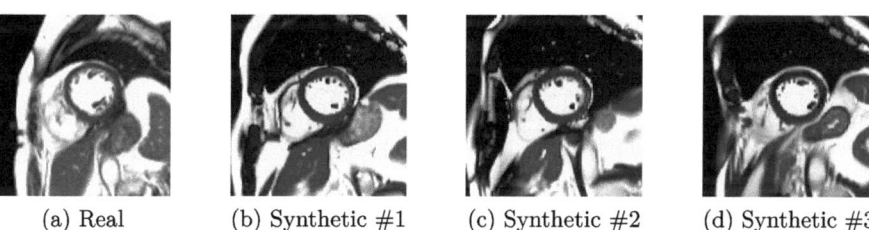

(a) Real (b) Synthetic #1 (c) Synthetic #2 (d) Synthetic #3

Fig. 5. Variability in generated CMR images using the same input but different seeds for the prompt: Female, 70s, overweight BMI, heart failure. Ventricle volumes (center) remain consistent, while realistic textural variations appear in surrounding anatomy. Figure 5a Reproduced by kind permission of UK Biobank 1'.

Qualitative Analysis. Sample images in Fig. 4 demonstrate the model's ability to link visual BMI indicators with textual prompts. Increased pericardial adipose tissue (PAT), marked with red arrows, is visible as BMI progresses. As noted in [19] PAT is a significant factor in discrimination of HF patients. Generated CMR images (Fig. 5) further showcase the model's ability to create a diverse set of samples and highlight the augmentation potential of the proposed approach.

3.2 Downstream Task: Heart Failure Classification

As presented in Table 1, integrating synthetic data with real samples led to an overall improvement in disease diagnosis performance, as reflected in higher AUROC and BACC scores, as well as improved average per-attribute scores. Specifically, the average BACC increased by 2% for groups separated by sex, 1.5% for BMI, and 1.3% for age. The ablation study in Fig. 6 further illustrates the impact of varying the proportion of synthetic data used during training. While the results exhibit a notable level of noise due to the limited number of test samples, a visible trend emerges – combining real and synthetic data provides a performance boost. On another note, while label-based weighting helps during training, subgroup weighting does not provide additional boost, likely due to much smaller subgroup sizes causing overfitting. From a fairness perspective, EOD improved for both sex and BMI attributes, though a slight decrease

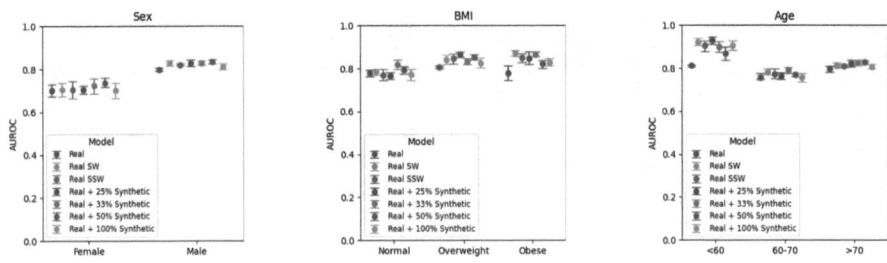

Fig. 6. Mean AUROC with 95% CI for each subgroup within the sensitive attributes.

Table 1. Average of per-group cardiac disease classification (CLF) scores for each sensitive attribute and overall performance for the whole population. $\text{CLF}_{Real+Synth}$ uses 33% synthetic data. Values multiplied by 100; best results in bold.

Group	Metric	CLF_{RealSW}	$\text{CLF}_{RealSSW}$	$\text{CLF}_{Real+Synth}$
Sex	AUROC ↑	78.9 ± 1.5	78.3 ± 1.7	**79.6 ± 1.6**
	BACC ↑	70.4 ± 1.2	68.4 ± 1.3	**72.4 ± 1.0**
	EOD ↓	37.2 ± 5.6	40.9 ± 4.7	**32.6 ± 6.1**
BMI	AUROC ↑	83.1 ± 1.3	82.2 ± 0.9	**83.8 ± 0.8**
	BACC ↑	73.7 ± 0.9	72.1 ± 1.2	**75.2 ± 0.7**
	EOD ↓	39.7 ± 5.7	37.8 ± 8.1	**32.1 ± 4.6**
Age	AUROC ↑	**83.7 ± 1.0**	82.8 ± 0.9	83.7 ± 0.9
	BACC ↑	74.9 ± 1.2	72.9 ± 1.5	**76.2 ± 0.8**
	EOD ↓	20.8 ± 6.1	**17.2 ± 3.6**	23.7 ± 7.5
Overall	AUROC ↑	83.1 ± 1.1	82.4 ± 0.9	**83.6 ± 0.8**
	BACC ↑	74.3 ± 0.9	72.7 ± 1.2	**75.8 ± 0.7**

was observed for age. These findings are consistent with the synthetic data quality analysis presented in 3.1. As shown in Fig. 6, females and individuals with a normal BMI experienced the most significant performance gains, effectively narrowing the gap to better-performing groups (e.g., males and obese individuals) by 13%. This aligns with the distribution imbalance observed in Fig. 1, where these populations had the lowest prevalence in the dataset, highlighting the potential of synthetic data to mitigate biases in model performance.

4 Discussion and Conclusion

In this work, we explore the use of generative latent diffusion models to address biases in CMR datasets. We show that combining textual (sex, age, BMI, heart condition) and imaging inputs (segmentation masks of cardiac shape) enables flexible and controllable synthetic data generation. Empirical evaluation on cardiac disease classification demonstrates performance gains for average per-group scores and fairness when training with synthetic balanced data, highlighting the potential of targeted data augmentation for reducing bias in cardiac imaging datasets. Our results also illustrate the challenge of addressing fairness in low-prevalence diseases, where subgroup sizes remain small and noisy even in large datasets. Future work on larger cohorts and more common diseases is needed to further assess this approach, including evaluation of subgroup-specific feature interpretability and analysis of changes in uncertainty estimates across subgroups. Additionally, we illustrate that such data augmentation is feasible on modest hardware, with all models—including multi-conditional diffusion models—trained on a single consumer-grade GPU, thereby laying the foundation for broader clinical adoption across diverse healthcare settings with varying computational resources.

Acknowledgment. This work was conducted using the UK Biobank resource under access application 2964. It received funding from the European Union's Horizon Europe research and innovation programme, Grant Agreement No. 101057849 (DataTools4Heart). NA acknowledges support from MRC Clinician Scientist Fellowship (MR/X020924/1). Barts Charity (G-002346) contributed to fees required to access UK Biobank data. This work acknowledges the support of the National Institute for Health and Care Research Barts Biomedical Research Centre (NIHR203330); a delivery partnership of Barts Health NHS Trust, Queen Mary University of London, St George's University Hospitals NHS Foundation Trust and St George's University of London. PG has received funding by the project TrustAI-ES (PID2023-146751OA-I00) from the Ministry of Science and Innovation of Spain.

Disclosure of Interests. SEP declares consultancy for Circle Cardiovascular Imaging, Inc., Calgary, Canada.

References

1. Deng, J., Dong, W., Socher, R., Li, L.J., Li, K., Fei-Fei, L.: Imagenet: a large-scale hierarchical image database. In: 2009 IEEE Conference on Computer Vision and Pattern Recognition, pp. 248–255 (2009)
2. He, K., Zhang, X., Ren, S., Sun, J.: Deep residual learning for image recognition. In: 2016 IEEE Conference on Computer Vision and Pattern Recognition (CVPR), pp. 770–778 (2016). https://doi.org/10.1109/CVPR.2016.90
3. Heusel, M., Ramsauer, H., Unterthiner, T., Nessler, B., Hochreiter, S.: Gans trained by a two time-scale update rule converge to a local nash equilibrium. In: Proceedings of the 31st International Conference on Neural Information Processing Systems, NIPS 2017, pp. 6629–6640. Curran Associates Inc., Red Hook, NY, USA (2017)
4. Ilharco, G., et al.: OpenCLIP (2021). https://doi.org/10.5281/zenodo.5143773
5. Jagannathan, R., Patel, S.A., Ali, M.K., Narayan, K.M.V.: Global Updates on Cardiovascular Disease Mortality Trends and Attribution of Traditional Risk Factors. Curr. Diab.Rep. **19**(7), 44 (2019). Jun
6. Konz, N., et al.: Fréchet radiomic distance (frd): A versatile metric for comparing medical imaging datasets (2025),. https://arxiv.org/abs/2412.01496

7. Ktena, I., et al.: Generative models improve fairness of medical classifiers under distribution shifts (2023)
8. Lambin, P., et al.: Radiomics: extracting more information from medical images using advanced feature analysis. Eur. J. Cancer **48**(4), 441–446 (2012)
9. Larrazabal, A.J., Nieto, N., Peterson, V., Milone, D.H., Ferrante, E.: Gender imbalance in medical imaging datasets produces biased classifiers for computer-aided diagnosis. Proc. Natl. Acad. Sci. **117**(23), 12592–12594 (2020). https://doi.org/10.1073/pnas.1919012117
10. Luo, L., Xu, D., Chen, H., Wong, T.T., Heng, P.A.: Pseudo bias-balanced learning for debiased chest x-ray classification. In: Wang, L., Dou, Q., Fletcher, P.T., Speidel, S., Li, S. (eds.) Medical Image Computing and Computer Assisted Intervention, pp. 621–631. Springer Nature Switzerland, Cham (2022). https://doi.org/10.1007/978-3-031-16452-1_59
11. Mikołajczyk, A., Majchrowska, S., Limeros, S.C.: The (de)biasing effect of GAN-based augmentation methods on skin lesion images (Jun 2022)
12. Osuala, R., et al.: Towards learning contrast kinetics with multi-condition latent diffusion models. In: International Conference on Medical Image Computing and Computer-Assisted Intervention, pp. 713–723. Springer (2024). https://doi.org/10.1007/978-3-031-72086-4_67
13. Paszke, A., et al.: Pytorch: an imperative style, high-performance deep learning library. In: Adv. Neural Inform. Process. Syst. **32**, 8024–8035 (2019)
14. Petersen, E., Feragen, A., Costa Zemsch, M.L., Henriksen, A., Wiese Christensen, O.E., Ganz, M.: Feature robustness and sex differences in medical imaging: A case study in mri-based alzheimer's disease detection. In: Wang, L., Dou, Q., Fletcher, P.T., Speidel, S., Li, S. (eds.) Medical Image Computing and Computer Assisted Intervention, pp. 88–98. Springer Nature Switzerland, Cham (2022). https://doi.org/10.1007/978-3-031-16431-6_9
15. Puyol-Antón, E., et al.: Fairness in cardiac magnetic resonance imaging: assessing sex and racial bias in deep learning-based segmentation. Front. Cardiovascular Med. **9** (2022)
16. Rombach, R., Blattmann, A., Lorenz, D., Esser, P., Ommer, B.: High-resolution image synthesis with latent diffusion models. In: Proceedings of the IEEE/CVF Conference on Computer Vision and Pattern Recognition (CVPR), pp. 10684–10695 (June 2022)
17. Schuhmann, C., et al.: LAION-5b: an open large-scale dataset for training next generation image-text models. In: Thirty-sixth Conference on Neural Information Processing Systems Datasets and Benchmarks Track (2022)
18. Sudlow, C., et al.: Uk biobank: an open access resource for identifying the causes of a wide range of complex diseases of middle and old age. PLoS Med. **12**(3), e1001779 (2015)
19. Szabo, L., et al.: Radiomics of pericardial fat: a new frontier in heart failure discrimination and prediction. European Radiology (Nov 2023)
20. Wachinger, C., Rieckmann, A., Pölsterl, S.: Detect and correct bias in multi-site neuroimaging datasets. Med. Image Anal. **67**, 101879 (2021)
21. Wang, Z., et alO.: Towards fairness in visual recognition: effective strategies for bias mitigation. In: 2020 IEEE/CVF Conference on Computer Vision and Pattern Recognition (CVPR), pp. 8916–8925 (2020)
22. Xing, X., Felder, F., Nan, Y., Papanastasiou, G., Simon, W., Yang, G.: You don't have to be perfect to be amazing: Unveil the utility of synthetic images. arXiv preprint arXiv:2305.18337 (2023)

23. Zare, S., Nguyen, H.V.: Removal of confounders via invariant risk minimization for medical diagnosis. In: Wang, L., Dou, Q., Fletcher, P.T., Speidel, S., Li, S. (eds.) Medical Image Computing and Computer Assisted Intervention, pp. 578–587. Springer Nature Switzerland, Cham (2022). https://doi.org/10.1007/978-3-031-16452-1_55
24. Zhang, B.H., Lemoine, B., Mitchell, M.: Mitigating unwanted biases with adversarial learning. In: Proceedings of the 2018 AAAI/ACM Conference on AI, Ethics, and Society, pp. 335–340. AIES 2018, Association for Computing Machinery, New York (2018). https://doi.org/10.1145/3278721.3278779
25. Zhang, L., Rao, A., Agrawala, M.: Adding conditional control to text-to-image diffusion models. In: IEEE International Conference on Computer Vision (ICCV) (2023)

Exploring the Interplay of Label Bias with Subgroup Size and Separability: A Case Study in Mammographic Density Classification

Emma A. M. Stanley[1,2,3,4(✉)], Raghav Mehta[5], Mélanie Roschewitz[5], Nils D. Forkert[2,3,4], and Ben Glocker[5]

[1] Department of Biomedical Engineering, University of Calgary, Calgary, Canada
emma.stanley@ucalgary.ca
[2] Department of Radiology, University of Calgary, Calgary, Canada
[3] Hotchkiss Brain Institute, University of Calgary, Calgary, Canada
[4] Alberta Children's Hospital Research Institute, University of Calgary, Calgary, Canada
[5] Department of Computing, Imperial College London, London, UK

Abstract. Systematic mislabelling affecting specific subgroups (*i.e.*, label bias) in medical imaging datasets represents an understudied issue concerning the fairness of medical AI systems. In this work, we investigated how size and separability of subgroups affected by label bias influence the learned features and performance of a deep learning model. Therefore, we trained deep learning models for binary tissue density classification using the EMory BrEast imaging Dataset (EMBED), where label bias affected separable subgroups (based on imaging manufacturer) or non-separable 'pseudo-subgroups'. We found that simulated subgroup label bias led to prominent shifts in the learned feature representations of the models. Importantly, these shifts within the feature space were dependent on both the relative size and the separability of the subgroup affected by label bias. We also observed notable differences in subgroup performance depending on whether a validation set with clean labels was used to define the classification threshold for the model. For instance, with label bias affecting the majority separable subgroup, the true positive rate for that subgroup fell from 0.898, when the validation set had clean labels, to 0.518, when the validation set had biased labels. Our work represents a key contribution toward understanding the consequences of label bias on subgroup fairness in medical imaging AI.

Keywords: Label Bias · Mammography · Fairness

1 Introduction

An implicit assumption when training artificial intelligence (AI) models for supervised classification tasks is that the labels provided with a training dataset represent the ground truth. However, the annotations for large-scale datasets, which are required for effectively training deep models, are costly to acquire, and may not be fully reliable due to factors such as human error, subjective or ambiguous labelling tasks, or inaccurate automated labelling systems [15]. Incorrectly labelled data—'label noise'—has been shown to negatively impact generalizability and training dynamics [2], comprising an often overlooked but potentially pervasive problem. Medical imaging datasets may be particularly susceptible to label noise, due to inter-annotator variability [12,17], the use of language models for automatically extracting labels from free-text radiology reports [6], and the inherent complexity of many imaging-based diagnostic tasks as primary contributing factors [14].

Label noise may be present throughout a dataset, but it may also systematically affect certain subsets of data. For instance, a low-quality scanner being used in an under-resourced setting [11] may produce images that are more challenging to identify lesions from. Annotations by a less experienced radiologist at a medical center that primarily serves a particular population demographic could result in a higher rate of diagnostic errors affecting that population [18]. Human decision-making biases could lead to higher rates of mis- or under-diagnosis for some genders or racial subgroups [10] within a dataset. A language model that extracts disease labels from radiology reports may perform inaccurately on reports written in different languages [8]. In any of the aforementioned examples, AI models trained on such datasets could learn an inconsistent mapping between image features and labels that systematically affects particular subpopulations, possibly leading to performance disparities between subgroups within the data.

While progress has been made toward identifying and addressing label noise in medical imaging datasets [14,16], little attention has been paid to the implications of systematic subgroup label noise (*i.e.*, label bias) as a potential fairness issue [13]. In this context, our work aims to investigate the impacts of subgroup label bias on a deep learning model, using tissue density classification from mammography data as an example application of clinical importance. Specifically, we (**1**) examine the feature space of a deep learning model trained under simulated subgroup label bias, illustrating how subgroup size and separability influence learned representations, and (**2**) demonstrate notable differences in subgroup classification performance depending on whether an unbiased validation set is used for defining the classification threshold.

2 Materials and Methods

2.1 Dataset

We investigated label bias in a breast tissue density classification task using full-field digital mammography images from the EMory BrEast imaging Dataset (EMBED), which were acquired by four institutional hospitals over a seven-year period [7]. The dataset was filtered to remove laterality mismatches, and images with spot compression and magnification. Only mediolateral oblique and craniocaudal views from patients identified as female who were assigned a single BI-RADS tissue density label of either A, B, C, or D were used. A density label of A corresponds to the lowest density (*i.e.*, mostly fatty tissue), while D corresponds to the highest density (*i.e.*, mostly dense tissue). For the classification task, we binarized the density labels into 0 := $\{A, B\}$ and 1 := $\{C, D\}$, and performed undersampling to balance the two classes. The dataset was further filtered to include only the three largest imaging system manufacturers: Hologic (HOLO), GE Medical Systems (GEMS), and Fujifilm (FUJI). HOLO comprised a large majority of the dataset, followed by GEMS and FUJI (see Table 1).

Table 1. Number of images belonging to each manufacturer subgroup.

Manufacturer	Total	Percent of Dataset	Tissue Density			
			A	B	C	D
Hologic (HOLO)	170,995	89.0	18,367	67,124	76,720	8,784
GE Medical Systems (GEMS)	15,965	8.3	636	7,210	7,590	529
Fujifilm (FUJI)	5,130	2.7	425	2,283	2,260	162

2.2 Subgroup Label Bias

We define subgroup label bias as the *systematic mislabelling of images affecting a single subgroup*. This was modelled by changing the binary tissue density labels for images in a subgroup (*e.g.*, from a particular manufacturer) from class 1 to class 0. However, we only introduced label bias to patients with tissue density category C, as we assumed this approach to be more realistic for potentially ambiguous cases (B/C) to be mislabelled, compared to categorical extremes (A/D). Furthermore, since many patients in the dataset had multiple images resulting from subsequent examinations, we assumed that label bias affected all images from a given patient. Therefore, in all experiments, subgroup label bias was represented as the change of binary class label 1 to 0 in 30% of patients who were assigned a tissue density of C within a given subgroup. We analyzed label bias applied to each of the three manufacturer subgroups, as well as in 'pseudo-subgroups' to evaluate the role of subgroup separability [9] in label bias scenarios (see Sect. 3.1).

2.3 Models and Training

A ResNet-18 [5] was used to classify images into the binary tissue density classes. The dataset was randomly split into 60%/20%/20% for training, validation, and testing, ensuring no patient overlap between splits. Images were preprocessed by masking out non-tissue regions and normalizing pixel values to the range [0,1]. Data augmentation for training included gamma correction, brightness/contrast jitter, and random affine transformations. Models were trained at a learning rate of 1×10^{-5} with class-balanced batch sampling. The model with the highest area under the receiver operating characteristic curve (AUC) on the validation set over 10 epochs was evaluated.

2.4 Feature Inspection

For the models trained on each label bias scenario, we performed an exploration of the learned features following the methods described in Glocker et al. [4]. Briefly described, the test set was passed through the trained model and the features from the penultimate layer were extracted. Dimensionality reduction was performed using principal component analysis (PCA). We visualized these features with kernel density estimation (KDE) plots of the first PCA mode (PC1), as it was most predictive of tissue density.

2.5 Classification Performance

We computed subgroup-specific true (TPR) and false positive rates (FPR) for the test set *using true (clean) class labels* at a threshold defined by 10% overall FPR on the validation set. Since validation data may also be affected by label bias in a real-world setting, we compared subgroup performance between thresholds selected using both clean labels and biased labels (as described in Sec. 2.2).

3 Results

3.1 Subgroup Separability

In this work, subgroup separability was defined as the ability of a deep learning model to detect which subgroup an image belongs to [9]. To determine subgroup separability, we trained a model using the same method described in Sec. 2.3 to classify subgroups. One-vs-rest AUC for each manufacturer subgroup (HOLO, GEMS, and FUJI) was 1.00, demonstrating that these subgroups were fully separable. To compare to the non-separable case, we created three equally-sized 'pseudo-subgroups', where subgroup labels were randomly assigned to each image in the dataset, corresponding to pseudo-subgroups 1, 2, and 3. A model trained to classify these subgroups achieved only chance-level AUC values of 0.50, indicating that these subgroups were not separable (*i.e.*, the model was not able to distinguish the three pseudo-subgroups from one another).

3.2 Impact of Subgroup Label Bias on Learned Features

Models were trained on the dataset with clean labels, representing the baseline, as well as on the datasets with label bias applied to pseudo-subgroup 1 (PS1) and to each manufacturer subgroup separately. KDE plots of the first principal component (PC1) of the learned features for the subgroups in each label bias scenario (PS1, FUJI, GEMS, and HOLO) are visualized in Fig. 1, with the corresponding subgroup feature distribution for the clean label baseline overlaid. Notably, although the model was only trained to predict binary density classes, the feature space along PC1 in the clean label scenario naturally organized into a relatively symmetric, ordinal pattern corresponding to tissue density categories A, B, C, and D.

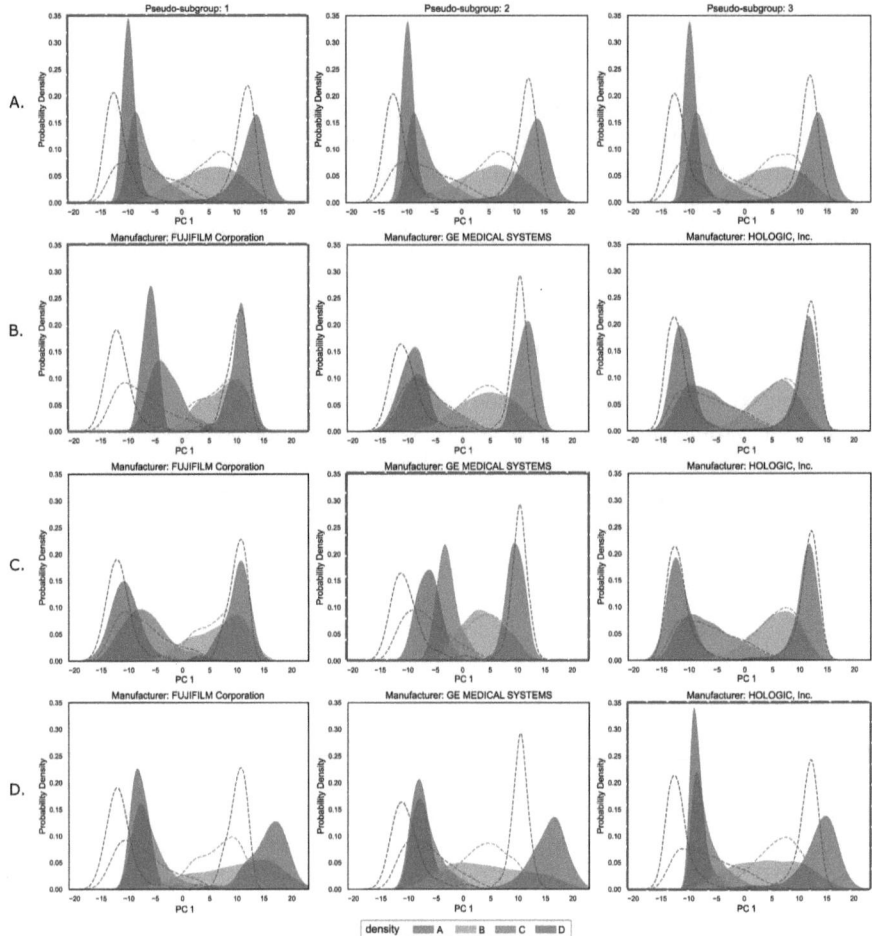

Fig. 1. Distribution of tissue density categories on the first principal component (PC1) of the feature space for each label bias scenario. Row A: label bias in pseudo-subgroup 1 (PS1), Row B: label bias in Fujifilm (FUJI), Row C: label bias in GE Medical Systems (GEMS), Row D: label bias in Hologic (HOLO). Dashed outlines show the distribution of tissue density categories for the specified subgroup from the model trained with clean labels. Red frames indicate the subgroup with label bias.

Non-separable Subgroups. When label bias was applied to density category C in PS1, the features for the entire class 1 (*i.e.*, density categories C and D) shifted and clustered more closely towards class 0. This can be seen in Fig. 1A as a higher concentration of the class 1 tissue density categories near the boundary separating the two classes, as compared to the clean label baseline. In this case, where label bias was applied to a non-separable subgroup (PS1), this feature shift was consistent across all pseudo-subgroups (PS1, PS2, PS3) (cf. Figure 1A subplots).

Separable Subgroups. Figs. 1B and 1C illustrate the impact of label bias applied to FUJI and GEMS respectively, both fully separable but minority subgroups. Here, it can be seen that the feature shift occurred primarily in the subgroup with label bias. The feature space of HOLO remained similar to the clean label baseline, and the feature space of the other minority subgroup experienced small perturbations but did not undergo a feature shift as prominent as the subgroup with label bias.

In contrast, when label bias was applied to HOLO, a separable subgroup that comprised a large majority of the dataset, the feature shift along PC1 of the class with label bias was evident across all subgroups (cf. Fig. 1D). In particular, while HOLO had the highest concentration of class 1 features near to class 0 (represented by the sharpest peak), GEMS and FUJI also had features for both C and D categories shift toward class 0. In all subgroups, the two class 1 density categories overlapped, as opposed to the ordinal pattern from the clean label baseline. It can also be seen that the distribution of class 0 tissue density categories had a wider spread, compared to the other label bias scenarios.

3.3 Impact of Subgroup Label Bias on Performance

TPR and FPR for the three manufacturer subgroups in each label bias scenario are presented in Table 2. Importantly, using a validation set with clean labels to define an operating point threshold led to different subgroup performance values compared to using a validation set with biased labels, depending on separability and size of the subgroup with label bias.

Table 2. True positive rate (TPR) and false positive rate (FPR) for each subgroup under each label bias scenario. Thresholds were computed at 10% overall FPR on the validation set, using either clean or biased labels. Values in parentheses indicate percent change relative to the clean baseline.

Subgroup with Label Bias	Validation Set Labels	TPR			FPR		
		Non-Separable Subgroups					
		PS1	PS2	PS3	PS1	PS2	PS3
Clean (Baseline)		0.916	0.913	0.915	0.106	0.104	0.104
PS1	Clean	0.910 (−00.7%)	0.912 (−00.1%)	0.912 (−00.3%)	0.106 (+00.0%)	0.108 (+03.8%)	0.104 (+00.0%)
PS1	Biased	0.826 (−09.8%)	0.828 (−09.3%)	0.826 (−08.6%)	0.042 (−60.4%)	0.045 (−56.7%)	0.048 (−53.8%)
		Separable Subgroups					
		FUJI	GEMS	HOLO	FUJI	GEMS	HOLO
Clean (Baseline)		0.923	0.914	0.915	0.069	0.170	0.100
FUJI	Clean	0.750 (−18.7%)	0.907 (−00.8%)	0.914 (−00.1%)	0.013 (−81.2%)	0.158 (−07.1%)	0.106 (+06.0%)
GEMS	Clean	0.941 (+02.0%)	0.780 (−14.7%)	0.914 (−00.1%)	0.092 (+33.3%)	0.087 (−48.8%)	0.103 (+03.0%)
HOLO	Clean	0.938 (+01.6%)	0.943 (−03.2%)	0.898 (−01.9%)	0.069 (+00.0%)	0.228 (+34.1%)	0.096 (−04.0%)
FUJI	Biased	0.728 (−21.1%)	0.904 (−01.1%)	0.910 (−00.5%)	0.013 (−81.2%)	0.153 (−10.0%)	0.102 (+02.0%)
GEMS	Biased	0.931 (+00.9%)	0.726 (−20.6%)	0.901 (−01.5%)	0.076 (+10.1%)	0.068 (−60.0%)	0.089 (−11.0%)
HOLO	Biased	0.844 (−08.6%)	0.828 (−09.4%)	0.518 (−43.4%)	0.015 (−78.3%)	0.080 (−52.9%)	0.012 (−88.0%)

Non-separable Subgroups. When clean labels were used for computing the threshold with label bias applied to PS1 during training, subgroup TPR and FPR for all subgroups were very similar to the clean label baseline. However, when biased labels were used for computing this threshold, TPR for all subgroups decreased by nearly 0.10, and FPR decreased by between 0.05–0.07.

Separable Subgroups. When label bias affected minority separable subgroups (*i.e.*, FUJI or GEMS), the TPR/FPR decreased for the affected subgroup, regardless of whether the threshold was set on a clean or biased-label validation set. For the subgroups without label bias, performance stayed relatively close to baseline, though usually slightly lower when the threshold was computed using the biased-label validation set. In contrast, the largest difference in subgroup performance between clean and biased validation sets resulted from label bias affecting HOLO, the majority separable subgroup. With a clean label validation set, TPR and FPR for HOLO decreased only slightly compared to the baseline. In contrast, when the classification threshold was computed using biased labels in the validation set, TPR for HOLO fell to 0.518, and TPR for GEMS and FUJI were also reduced by around 0.10 compared to when clean labels were used. FPR for all separable subgroups also decreased substantially.

4 Discussion

We demonstrated that the impact of subgroup label bias in a deep learning model trained for a binary classification task depended on both separability and relative size of the affected subgroup. In general, label bias led to a feature shift along PC1, in which features for the class affected by label bias shifted toward the other class. When label bias affected a non-separable subgroup, this feature shift occurred across all other non-separable subgroups. In contrast, when label bias was applied to a separable minority subgroup, this feature shift was primarily evident within the affected subgroup, possibly because the model learned that this 'noisy' mapping of images to labels was only associated with features of that particular subgroup. However, with label bias applied to the majority separable subgroup, all subgroups experienced a feature shift, similar to the non-separable case. This may be due to the much higher frequency that noisy labels were observed during training in this case, effectively causing the model to associate this noisy mapping with the entire dataset.

We speculate that the feature shifts observed in PC1 likely also led to a shift of optimal class separation thresholds in the high-dimensional decision space of the model. Figure 2 contains a simple toy example that illustrates how feature and threshold shifts caused by label bias could help to explain the impacts on subgroup performance presented in Table 2.

For example, consider label bias affecting a non-separable subgroup (Fig. 2B). In this case, selecting an operating point threshold using the clean labels would still lead to high classification performance, even though the feature space (and the corresponding optimal class decision threshold) would have shifted. However, using a validation set with biased labels would move the selected operating point threshold toward class 1, since calculation of the overall FPR would take

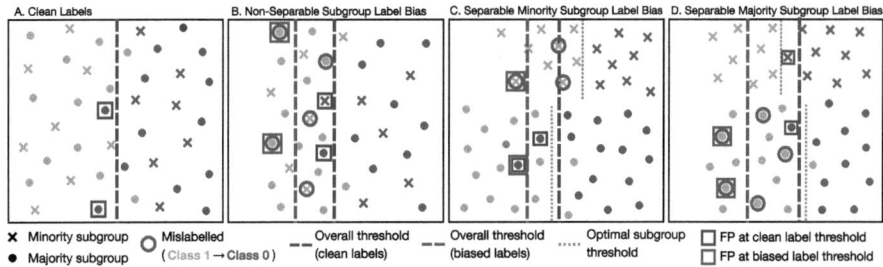

Fig. 2. Toy example illustrating how feature shifts caused by label bias could impact subgroup performance, when the operating point threshold of the model is selected using overall FPR on the validation set.

mislabelled images into account. This would result in lower TPR and FPR for all subgroups, similar to what we observed when label bias affected PS1.

With label bias in a minority separable subgroup (Fig. 2C), we observed a feature shift predominantly in the affected subgroup. However, since the classification threshold is based on overall FPR, this threshold would be primarily determined by the majority subgroup. Therefore, with a clean validation set, TPR and FPR for the minority subgroup would decrease since the optimal threshold for that subgroup would be shifted toward class 0, compared to the overall threshold. Using biased labels for validation, the overall threshold would be shifted slightly more toward class 1, leading to further decreased TPR and FPR for all subgroups, as we observed when label bias affected GEMS and FUJI.

In the case of label bias affecting the majority subgroup (Fig 2D), the classification threshold would be shifted toward class 0, but performance for the majority group would remain similar to the baseline. Assuming the optimal threshold for the majority subgroup shifted more than for the minority subgroup, TPR and FPR for the minority groups would increase. On the other hand, using biased labels for the overall FPR calculation would move the operating point threshold much further in the class 1 direction to account for the mislabelled images in the majority subgroup. This would then lead to a drastic decrease in TPR and FPR, particularly for the majority subgroup, as we observed when label bias was applied to HOLO.

5 Conclusions

The impact of subgroup label bias in medical imaging datasets is under-explored, but has important implications in the context of AI fairness. It has been shown that sociodemographic subgroups in medical imaging datasets can have a range of separability, for example, sex in retinal fundus images may have relatively low separability [9], while self-reported race subgroups in chest x-rays can be nearly completely separable [3]. Furthermore, substantial differences in relative subgroup size are often present, particularly within self-reported race categories [1].

Our results indicate that subgroup separability and size have distinct impacts on group fairness under label bias and are, therefore, both crucial to consider in future investigations. From a clinical usability standpoint, where it is necessary to define classification thresholds, our work emphasizes the importance of understanding whether datasets used for validating models are affected by label bias. In particular, defining an operating point using biased labels could substantially impact performance for one or all subgroups, if the affected subgroup is a large majority or if subgroups are non-separable. Therefore, future work should not only further evaluate the impacts of label bias in controlled scenarios, but also continue to work toward developing methods for reliably detecting the presence of label bias in medical imaging datasets.

Limitations. It is important to note that we assumed that the tissue density labels in the dataset were initially clean (*i.e.,* that there was no significant existing label bias). Furthermore, it is possible that other separable subgroups existed in the dataset, which may have introduced interacting effects. We explored only the first principal component of the feature space, therefore, confirming that the trends illustrated in the toy example correspond to the high-dimensional decision space of the model is a next step in future work. We studied separable subgroups defined by hardware-related differences, and only considered a static amount (30%) of label bias in one direction (*e.g.,* class 1 to class 0). Future work should investigate varying degrees of label bias affecting both classes, and whether the observed trends hold for label bias in sociodemographic subgroups.

Reproducibility. Code for data processing, experiments, and analysis is available at https://github.com/biomedia-mira/mammo-label-bias.

Acknowledgment. E.A.M.S. and N.D.F. acknowledge support from the Natural Science and Engineering Research Council of Canada, the Killam Trusts, Alberta Innovates, the Canada Research Chairs Program, and the River Fund at Calgary Foundation. M.R. is funded by an Imperial College London President's PhD Scholarship and a Google PhD Fellowship. R.M. is funded by the European Union's Horizon Europe research and innovation programme under grant agreement 101080302. B.G. received support from the Royal Academy of Engineering as part of his Kheiron/RAEng Research Chair.

Disclosure of Interests. B.G. is part-time employee of DeepHealth. No other competing interests..

References

1. Chen, R.J., et al.: Algorithmic fairness in artificial intelligence for medicine and healthcare. Nat. Biomed. Eng. **7**(6), 719–742 (2023)
2. Frenay, B., Verleysen, M.: Classification in the presence of label noise: a survey. IEEE Trans. Neural Netw. Learn. Syst. **25**(5), 845–869 (2014). May
3. Gichoya, J.W., et al.: Ai recognition of patient race in medical imaging: a modelling study. Lancet Digital Health **4**(6), e406–e414 (2022)
4. Glocker, B., Jones, C., Bernhardt, M., Winzeck, S.: Algorithmic encoding of protected characteristics in chest x-ray disease detection models. eBioMedicine **89**, 104467 (2023)

5. He, K., Zhang, X., Ren, S., Sun, J.: Deep residual learning for image recognition. In: 2016 IEEE Conference on Computer Vision and Pattern Recognition (CVPR), pp. 770–778 (Jun 2016), iSSN: 1063-6919
6. Jain, S., et al.: VisualCheXbert: addressing the discrepancy between radiology report labels and image labels. In: Proceedings of the Conference on Health, Inference, and Learning, CHIL 2021, pp. 105–115. Association for Computing Machinery, New York (Apr 2021)
7. Jeong, J.J., et al.: The emory breast imaging dataset (embed): a racially diverse, granular dataset of 3.4 million screening and diagnostic mammographic images. Radiology: Artifi. Intell. **5**(1), e220047 (2023)
8. Jones, C., Castro, D.C., De Sousa Ribeiro, F., Oktay, O., McCradden, M., Glocker, B.: A causal perspective on dataset bias in machine learning for medical imaging. Nat. Mach. Intell., 1–9 (2024)
9. Jones, C., Roschewitz, M., Glocker, B.: The role of subgroup separability in group-fair medical image classification. In: Greenspan, H., et al. (eds.) Medical Image Computing and Computer Assisted Intervention – MICCAI 2023, pp. 179–188. Lecture Notes in Computer Science, Springer Nature Switzerland, Cham (2023). https://doi.org/10.1007/978-3-031-43898-1_18
10. Markowitz, D.M.: Gender and ethnicity bias in medicine: a text analysis of 1.8 million critical care records. PNAS Nexus **1**(4), pgac157 (2022)
11. Mollura, D.J., et al.: Artificial intelligence in low- and middle-income countries: innovating global health radiology. Radiology **297**(3), 513–520 (2020)
12. Nichyporuk, B., et al.: Rethinking generalization: the impact of annotation style on medical image segmentation. Mach Learn. Biomed. Imaging **1**, 1–37 (2022)
13. Petersen, E., Holm, S., Ganz, M., Feragen, A.: The path toward equal performance in medical machine learning. Patterns **4**(7), 100790 (2023). Jul
14. Shi, J., Zhang, K., Guo, C., Yang, Y., Xu, Y., Wu, J.: A survey of label-noise deep learning for medical image analysis. Med. Image Anal. **95**, 103166 (2024). Jul
15. Xiao, T., Xia, T., Yang, Y., Huang, C., Wang, X.: Learning from massive noisy labeled data for image classification. In: 2015 IEEE Conference on Computer Vision and Pattern Recognition (CVPR), pp. 2691–2699. IEEE, Boston, MA, USA (Jun 2015)
16. Wei, Y., Deng, Y., Sun, C., Lin, M., Jiang, H., Peng, Y.: Deep learning with noisy labels in medical prediction problems: a scoping review. J. Am. Med. Inform. Assoc. **31**(7), 1596–1607 (2024). Jul
17. Yang, F., et al.: Assessing inter-annotator agreement for medical image segmentation. IEEE access: practical innovations, open solutions **11**, 21300–21312 (2023)
18. Zhang, L., Wen, X., Li, J.W., Jiang, X., Yang, X.F., Li, M.: Diagnostic error and bias in the department of radiology: a pictorial essay. Insights Imaging **14**, 163 (2023). Oct

Does a Rising Tide Lift All Boats? Bias Mitigation for AI-Based CMR Segmentation

Tiarna Lee[1(✉)], Esther Puyol-Antón[1,4], Bram Ruijsink[1,2], Miaojing Shi[3], and Andrew P. King[1]

[1] School of Biomedical Engineering and Imaging Sciences, King's College London, London, UK
tiarna.lee@kcl.ac.uk
[2] Guy's and St Thomas' Hospital, London, UK
[3] College of Electronic and Information Engineering, Tongji University, Shanghai, China
[4] HeartFlow Inc., London, UK

Abstract. Artificial intelligence (AI) is increasingly being used for medical imaging tasks. However, there can be biases in the resulting models, particularly when they were trained using imbalanced training datasets. One such example has been the strong race bias effect in cardiac magnetic resonance (CMR) image segmentation models. Although this phenomenon has been reported in a number of publications, little is known about the effectiveness of bias mitigation algorithms in this domain. We aim to investigate the impact of common bias mitigation methods to address bias between Black and White subjects in AI-based CMR segmentation models. Specifically, we use oversampling, importance reweighing and Group DRO as well as combinations of these techniques to mitigate the race bias. Furthermore, motivated by recent findings on the root causes of AI-based CMR segmentation bias, we evaluate the same methods using models trained and evaluated on cropped CMR images. We find that bias can be mitigated using oversampling, significantly improving performance for the underrepresented Black subjects whilst not significantly reducing the majority White subjects' performance. Group DRO also improves performance for Black subjects but not significantly, while reweighing decreases performance for Black subjects. Using a combination of oversampling and Group DRO also improves performance for Black subjects but not significantly. Using cropped images increases performance for both races and reduces the bias, whilst adding oversampling as a bias mitigation technique with cropped images reduces the bias further.

Keywords: Bias mitigation · CMR imaging · Segmentation

1 Introduction

Artificial intelligence (AI) is increasingly being used to aid medical diagnosis, prognosis and treatment planning. However, AI models have been shown to exhibit bias by protected attributes in many different applications. For example, in cardiac magnetic resonance (CMR) image segmentation, both sex [7] and race [5,9,10] bias have been reported. Sex bias has also been found in chest X-ray image classification [4,12]. AI bias can have detrimental downstream impacts in medical imaging applications. For example, CMR segmentations are used to derive biomarkers whose values impact patient management, so greater errors in these biomarkers for certain protected groups can lead to inappropriate treatment choices and worse outcomes [9].

Previous work has aimed to address biases in AI models for medical imaging tasks by using bias mitigation methods. [14] evaluated and mitigated bias in chest X-ray classification models using methods based on oversampling minority protected groups and Pareto minimax optimisation. [15] proposed a framework with eleven algorithms which aimed to measure and mitigate biases in medical imaging classification datasets. [13] evaluated stratified batch sampling, a balanced dataset model and a protected group-specific model for orthopaedic image segmentation. In CMR segmentation, [10] proposed oversampling of minority protected groups and a multi-task learning approach which learnt both a segmentation model and a protected attribute classifier.

Recently, an investigation was performed to discover the root cause(s) of AI-based CMR segmentation bias [6]. One key finding of this work was that distributional differences in areas outside the heart were significant in the learning of biased representations in AI segmentation models. This finding opens up new avenues of enquiry in bias mitigation, which we explore in this paper.

The contribution of this paper is to perform the most extensive and comprehensive investigation to date of bias mitigation methods in AI-based CMR segmentation. We investigate multiple mitigation methods as well as combinations of these methods. We also investigate AI models trained using CMR images that are cropped to remove features outside of the heart that may be leading to bias, and evaluate the effectiveness of bias mitigation techniques in this setting. Our results show that CMR segmentation bias can be mitigated to produce results that have high accuracy but are not significantly different between races.

2 Methods

We use CMR images from the UK Biobank [8]. The dataset consists of end diastolic (ED) and end systolic (ES) cine short-axis images from 5,778 subjects. Manual segmentation of the left ventricular blood pool (LVBP), left ventricular myocardium (LVM), and right ventricular blood pool (RVBP) was performed for the ED and ES images of each subject. The LV endocardial and epicardial borders and the RV endocardial border were outlined using cvi42 (version 5.1.1, Circle Cardiovascular Imaging Inc., Calgary, Alberta, Canada). The same guidelines were provided to a panel of ten experts with one expert annotating each image. Each expert was provided with a random selection of images for annotation which included subjects of different sexes and races. They were not provided with demographic information about the subjects.

Previous work [5,7] has shown that bias is greater when the imbalance between races in the training set is greater. Therefore, we curate a dataset where biases would be significant to allow for better evaluation of mitigation methods. Our training set comprises 15 Black subjects and 4,221 White subjects.

2.1 Baseline Model

We trained a baseline 2D nnU-Net v1 model [2] to segment the LVM, LVBP and RVBP using the training parameters from [7]. The demographic information of the subjects used in the training and testing datasets can be seen in Table 1 and Table 2. Training was performed using only Black and White subjects but testing was performed using all other available subjects including Mixed, Asian, Chinese and Other.

Table 1. Characteristics of subjects used in the training dataset. Mean (standard deviation) values are presented for each characteristic. Statistically significant differences between subject groups and the overall average are indicated with an asterisk * ($p < 0.05$) and were determined using a two-tailed Student's t-test.

Health measure	Overall	White	Black
# subjects	4236	4221	15
Age (years)	64.6 (7.7)	64.6 (7.7)	57.0 (4.9)*
Weight (kg)	77.0 (15.0)	77.0 (15.0)	74.8 (11.9)*
Standing height (cm)	169.5 (9.2)	169.5 (9.2)	168.8 (6.9)*
Body Mass Index	26.7 (4.3)	26.7 (4.3)	26.2 (3.7)

Table 2. Characteristics of subjects used in the test dataset. Mean (standard deviation) values are presented for each characteristic. Statistically significant differences between subject groups and the overall average are indicated with an asterisk * ($p < 0.05$) and were determined using a two-tailed Student's t-test.

Health measure	Overall	White	Mixed	Asian	Black	Chinese	Other
# subjects	1542	469	170	387	223	111	182
Age	61.7 (7.9)	64.8 (7.7)*	59.8 (7.2)*	61.0 (8.2)*	59.1 (7.1)*	59.7 (6.4)*	62.1 (7.5)
Standing height	167.4 (9.2)	169.8 (9.4)*	166.4 (8.5)	166.7 (8.6)	168.7 (9.4)	162.4 (7.1)*	165.4 (9.7)*
Weight	75.1 (15.1)	77.7 (15.1)*	75.1 (15.3)	72.3 (12.8)*	82.0 (16.0)*	63.5 (9.9)*	73.3 (15.5)
Body Mass Index	26.7 (4.4)	26.9 (4.2)	27.1 (5.2)	25.9 (3.7)*	28.8 (5.1)*	24.0 (3.1)*	26.6 (4.4)

2.2 Oversampling

We also trained a 2D nnU-Net model using the same data as the baseline but applied oversampling during batch selection [3]. Oversampling refers to the process of increasing the sampling of a minority group in a dataset. Here, we oversample the Black subjects in the training dataset so that they were equal to the number of White subjects in each batch used during training. This was performed using random sampling with replacement so each subject could in principle be selected more than once in a training batch. The same training parameters from [7] were used.

2.3 Reweighing

An nnU-Net model was also trained using a reweighing mitigation strategy [3]. Reweighing refers to the process of increasing the importance of under-represented groups to the model. This can be done as a pre-processing step or during training by adapting the loss function. We performed this during training by adding a weighting term to the combined Cross Entropy (CE)-Dice loss function of the nnU-Net. Each group was weighted inversely proportionally to the group size, as shown in Eq. 1. The weights were then normalised so that they summed to 1. The training parameters from [7] were again used.

$$\text{Weights per group:} \quad w_g = \frac{n_G}{n_g + \epsilon}, \quad \text{where } \epsilon = 10^{-6}. \tag{1}$$

$$\text{Normalized weights:} \quad \hat{w}_g = \frac{w_g}{\sum_{j=1}^{N_G} w_j}, \quad g = 1, 2, \ldots, N_G. \tag{2}$$

where n_g is the number of samples in protected group g, n_G is the number of samples in all groups and N_G is the number of groups.

2.4 Group Distributionally Robust Optimisation

The final mitigation approach was Group Distributionally Robust Optimisation, or Group DRO, which was first proposed in [11]. The method aims to optimise the performance of the worst-performing group in a dataset. The Group DRO loss function can be formalised as:

$$L_{DRO} = max_{g \in G} \frac{1}{n_g} \sum_{i \in g} L(y_i, \hat{y}_i) \tag{3}$$

where L is the loss function computed between predicted labels $\hat{y}_i$ and ground truth labels y_i.

In this method, CE loss was used instead of CE-Dice loss which was used for the oversampling and reweighing experiments. Group DRO uses losses from individual samples to calculate the average loss for the groups. However, Dice loss is calculated using global statistics of the true positives, false positives and false negatives for a group or batch. It is non-additive as the numerator and denominator will change if the calculation is performed on a per-sample basis rather than for a group or batch. For Dice loss, the average group loss and global loss are different, which causes instability in training.

2.5 Training Using Cropped Images

Following the findings of [6], which found that areas outside the heart were a contributing factor to CMR segmentation performance bias, we also performed experiments using all of the above techniques for a nnU-Net model trained using cropped CMR images. The images were cropped around the heart using a bounding box defined based on the ground truth segmentation. All images were cropped to the same size, i.e. the size of the largest heart in the dataset.

2.6 Training Using Combinations of Mitigation Methods

The final experiment combined the mitigation methods into pairs to test whether combinations of methods would improve performance. This results in three additional methods: oversampling + Group DRO, reweighing + Group DRO, and oversampling + reweighing. These methods were applied in the same way as above. CE loss was used for experiments with Group DRO and CE-Dice loss was used for oversampling + reweighing.

2.7 Metrics

For each of the methods described above, performance was measured by finding the overall Dice similarity coefficient (DSC) for subjects in the test set. Performance was measured using the median and inter-quartile range (IQR) of the DSC. Fairness metrics are also reported: the fairness gap (FG) as defined in [7] and skewed error ratio (SER) as defined in [10]. The SER is defined as $SER = \frac{max_g(1-D_g)}{min_g(1-D_g)}$ where D_g is the median DSC for protected group g. This measures the ratio of the errors between the median DSCs for the worst-performing and best-performing protected groups. A more fair model will have a FG closer to 0 and a SER closer to 1.

3 Results

The results of the baseline and the three bias mitigation methods can be seen in Table 3 and in Fig. 1. Oversampling was the only method to increase performance for the Black subjects such that there was no significant difference between the median DSC scores of the Black and White subjects. The method increased the median performance for Black subjects by 0.045. Fairness performance metrics (SER and FG) also decreased, showing more equitable performance. Using oversampling also caused performance to increase for the other races compared to the baseline, as can be seen in Fig. 1a and Fig. 1b. This is perhaps surprising as performance on these other races was not optimised during training.

The other mitigation methods did not significantly improve performance for the Black subjects. Reweighing resulted in worse performance for the Black subjects, with the median DSC decreasing. Group DRO resulted in increased performance for the Black subjects but performance remained significantly lower than for the White subjects. Both oversampling and Group DRO slightly decreased median DSC for White subjects (although not statistically significantly), but reweighing significantly decreased median DSC for White subjects.

Table 3. DSC values for each of the bias mitigation methods. The p-values were computed between White and Black subjects based on a two-sided Mann Whitney U test. * p< 0.0001. The best median DSC score for Black and White subjects is shown in bold

	Baseline *		Oversampling p = 0.22		Reweighing *		Group DRO *	
	White	Black	White	Black	White	Black	White	Black
Median	**0.896**	0.846	0.894	**0.891**	0.891	0.831	0.893	0.865
IQR	0.046	0.064	0.045	0.049	0.049	0.069	0.047	0.051
SER	1.486		1.032		1.544		1.260	
Fairness gap	0.050		0.003		0.059		0.028	

As shown in Table 4, the baseline model trained using cropped images improved performance for both White and Black subjects and reduced bias. Oversampling significantly improved performance for the Black subjects compared to the baseline, resulting in performance that was higher than for White subjects. Group DRO improved the DSC for Black subjects but this increase was not significant. Reweighing made performance slightly worse for both races, with median DSC decreasing and SER and FG measures increasing.

Table 4. DSC values for each of the bias mitigation methods using cropped images. The p-values were computed between White and Black subjects based on a two-sided Mann Whitney U test. * p< 0.0001. The best median DSC score for Black and White subjects is shown in bold

	Baseline *		Oversampling*		Reweighing *		Group DRO *	
	White	Black	White	Black	White	Black	White	Black
Median	**0.919**	0.893	0.913	**0.922**	0.916	0.887	0.918	0.908
IQR	0.032	0.041	0.030	0.039	0.033	0.043	0.033	0.038
SER	1.326		1.115		1.352		1.130	
Fairness gap	0.026		−0.009		0.029		0.011	

Finally, combining the mitigation methods did not produce significantly less biased results, as shown in Table 5. All three combinations decreased perfor-

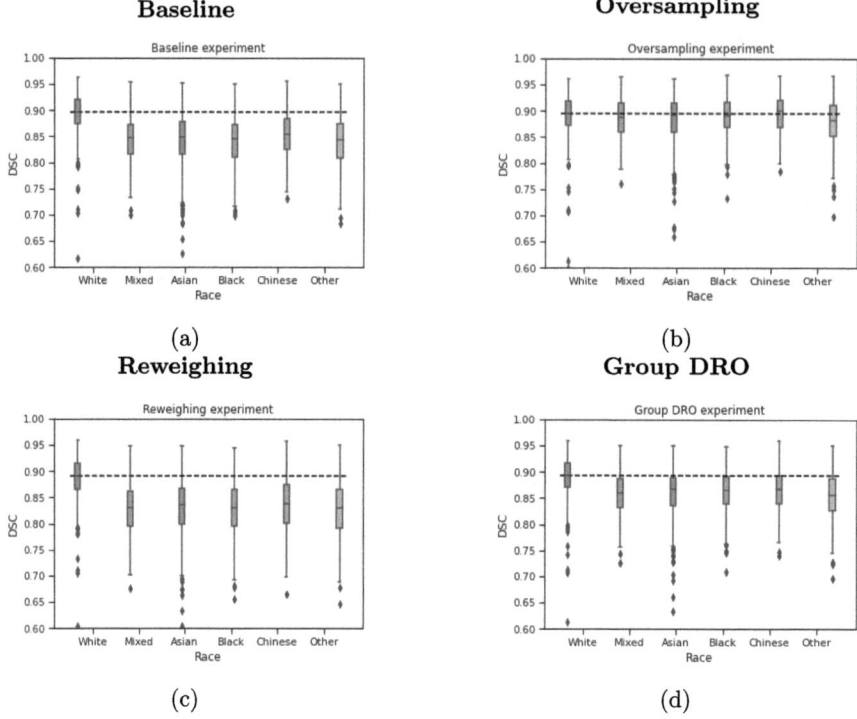

Fig. 1. Overall DSC for bias mitigation methods on uncropped images. The dashed line indicates median DSC for White test subjects

mance for White subjects but increased performance for Black subjects. The best combination was oversampling and Group DRO which reduced the performance for the White subjects the least and improved performance for the Black subjects the most, leading to the lowest FG.

4 Discussion

This work has performed a comprehensive examination of bias mitigation methods for AI segmentation models used for cine CMR images. We have shown that bias in CMR segmentation models can be mitigated by using such methods. In particular, oversampling minority subjects reduces bias so that there is no significant difference between the performance of the Black and White subjects. Although oversampling did not add any extra information to the dataset, the method allowed the network to train on Black subjects more frequently than if oversampling was not used, allowing for better balance between protected groups. It could be anticipated that training using a small number of (oversampled) Black subjects would increase the risk of overfitting to those subjects, having a detrimental effect on generalisation, but this effect was not seen in our

Table 5. DSC values for the combined bias mitigation methods using original sized images. The p-values were computed between White and Black subjects based on a two-sided Mann Whitney U test. * p< 0.0001. The best median DSC score for Black and White subjects is shown in bold

	Baseline *		Oversampling + Group DRO *		Group DRO + Reweighing *		Oversampling + Reweighing *	
	White	Black	White	Black	White	Black	White	Black
Median	**0.896**	0.846	0.889	**0.881**	0.893	0.863	0.889	0.871
IQR	0.046	0.064	0.048	0.054	0.047	0.055	0.051	0.057
SER	1.486		1.08		1.285		1.161	
Fairness gap	0.050		0.009		0.031		0.017	

experiments as test performance remained high. Previous work in [6] showed that race could be classified from cine CMR images, indicating that there are distinct features in the images of different races that are recognisable to AI models. Using oversampling will allow the network to see more of these distinct features to learn better representations of the under-represented group.

Reweighing did not improve segmentation performance, instead decreasing performance for both protected groups. This may be due to increased importance being given to a small group of subjects, decreasing focus on the larger group. As described in [5] and [7], when White subjects comprised 75% of the training set, their segmentation performance was still lower than for the Black subjects who comprised 25% of the training set. This suggests that segmentation of the White subjects' images may be a more difficult task as there may be more outliers and variation in the hearts than in the Black subjects. Reweighing these White subjects so that their importance is lower in the loss function could decrease accuracy.

Using cropped images increased performance and reduced bias compared to using uncropped images. Using oversampling with cropped images reduced bias further, but interestingly not to the same extent as using uncropped images. Also, note that to use a model trained using cropped images in clinical practice, a region-of-interest proposal network would be required to identify the cropping region at test time [1].

The combination of Group DRO and oversampling produced a model which improved performance for Black subjects but was still significantly lower than for White subjects. Therefore, using a combined method was better than using Group DRO alone but worse than oversampling alone.

This work has some limitations. For example, only Black and White subjects were used during training to control for other factors which may affect mitigation methods. Future work could investigate the effect of mitigating biases using multiple races or in other protected attributes such as age and socioeconomic status. It would also be useful to investigate the performance of bias mitigation strategies under external validation.

Acknowledgements. This work was supported by the Engineering & Physical Sciences Research Council Doctoral Training Partnership (EPSRC DTP) grant EP/T517963/1. This research has been conducted using the UK Biobank Resource under Application Number 17806.

References

1. He, K., et al.: Mask R-CNN. IEEE Trans. Pattern Anal. Mach. Intell. **42**(2), 386–397 (2017). issn: 19393539. https://doi.org/10.1109/TPAMI.2018.2844175. https://arxiv.org/abs/1703.06870v3
2. Isensee, F., et al.: nnU-Net: a self-configuring method for deep learning based biomedical image segmentation. Nat. Methods **18**(2), 203–211 (2020). issn: 1548-7105. https://doi.org/10.1038/s41592-020-01008-z
3. Kamiran, F., Calders, T.: Data preprocessing techniques for classification without discrimination. Knowl. Inf. Syst. **33**(1), 1–33 (2012). issn: 02193116. https://doi.org/10.1007/S10115-011-0463-8/METRICS. https://link.springer.com/article/10.1007/s10115-011-0463-8
4. Larrazabal, A.J., et al.: Gender imbalance in medical imaging datasets produces biased classifiers for computer-aided diagnosis. Proc. Natl. Acad. Sci. USA **117**(23), 12592–12594 (2020). issn: 1091-6490. https://doi.org/10.1073/pnas.1919012117
5. Lee, T., et al.: A systematic study of race and sex bias in CNN based cardiac MR segmentation. In: Lecture Notes in Computer Science (including subseries Lecture Notes in Artificial Intelligence and Lecture Notes in Bioinformatics). LNCS, vol. 13593, pp. 233–244. Springer, Heidelberg (2022). isbn: 9783031234422. https://doi.org/10.1007/978-3-031-23443-9_22
6. Lee, T., et al.: An investigation into the causes of race bias in AI-based cine CMR segmentation. Eur. Heart J. Dig. Health (2025). issn: 2634-3916. https://doi.org/10.1093/EHJDH/ZTAF008
7. Lee, T., et al.: An investigation into the impact of deep learning model choice on sex and race bias in cardiac MR segmentation. In: Lecture Notes in Computer Science (including subseries Lecture Notes in Artificial Intelligence and Lecture Notes in Bioinformatics). LNCS, vol. 14242, pp. 215–224 (2023). issn: 16113349. https://doi.org/10.1007/978-3-031-45249-9_21/FIGURES/2. https://link.springer.com/chapter/10.1007/978-3-031-45249-9_21
8. Petersen, S.E., et al.: UK Biobank's cardiovascular magnetic resonance protocol. J. Cardiovasc. Magn. Reson. **18**(1), 1–7 (2016). issn: 1532429X. https://doi.org/10.1186/s12968-016-0227-4
9. Puyol-Antón, E., et al.: Fairness in cardiac magnetic resonance imaging: assessing sex and racial bias in deep learning-based segmentation. Front. Cardiovasc. Med. 664 (2022). issn: 2297-055X. https://doi.org/10.3389/FCVM.2022.859310
10. Puyol-Antón, E., et al.: Fairness in cardiac MR image analysis: an investigation of bias due to data imbalance in deep learning based segmentation. In: Medical Image Computing and Computer Assisted Intervention – MICCAI 2021. LNCS, vol. 12903, pp. 413–423. Springer, Heidelberg (2021). isbn: 9783030871987. https://doi.org/10.1007/978-3-030-87199-4_39
11. Sagawa, S., et al.: Distributionally robust neural networks for group shifts: on the importance of regularization for worst-case generalization. In: 8th International Conference on Learning Representations, ICLR 2020 (2020)

12. Seyyed-Kalantari, L., et al.: CheXclusion: fairness gaps in deep chest X-ray classifiers. In: Pacific Symposium on Biocomputing. Pacific Symposium on Biocomputing, vol. 26, pp. 232–243 (2021). issn: 23356936. https://doi.org/10.1142/9789811232701_0022
13. Siddiqui, I.A., et al.: Fair AI-powered orthopedic image segmentation: addressing bias and promoting equitable healthcare. Sci. Rep. **14**(123), 16105 (2024). https://doi.org/10.1038/s41598-024-66873-6
14. Zhang, H., et al.: Improving the fairness of chest X-ray classifiers, p. 2022 (2022). https://doi.org/10.48550/arxiv.2203.12609. https://arxiv.org/abs/2203.12609v1
15. Zong, Y., Yang, Y., Hospedales, T.: MEDFAIR: benchmarking fairness for medical imaging (2022). https://doi.org/10.48550/arxiv.2210.1725. https://arxiv.org/abs/2210.01725v1

MIMM-X: Disentangling Spurious Correlations for Medical Image Analysis

Louisa Fay[1,2(✉)], Hajer Reguigui[2], Bin Yang[2], Sergios Gatidis[3], and Thomas Küstner[1]

[1] Medical Image and Data Analysis, University Hospital of Tübingen, Tübingen, Germany
louisa.fay@med.uni-tuebingen.de
[2] Institute for Signal Processing and System Theory, University of Stuttgart, Stuttgart, Germany
[3] Department of Radiology, Stanford University, Stanford, USA

Abstract. Deep learning models can excel on medical tasks, yet often experience spurious correlations, known as shortcut learning, leading to poor generalization in new environments. Particularly in medical imaging, where multiple spurious correlations can coexist, misclassifications can have severe consequences. We propose MIMM-X, a framework that disentangles causal features from multiple spurious correlations by minimizing their mutual information. It enables predictions based on true underlying causal relationships rather than dataset-specific shortcuts. We evaluate MIMM-X on three datasets (UK Biobank, NAKO, CheXpert) across two imaging modalities (MRI and X-ray). Results demonstrate that MIMM-X effectively mitigates shortcut learning of multiple spurious correlations. The code is publicly available https://github.com/lab-midas/MIMM-X.

Keywords: Causality · Disentanglement · Shortcut Learning

1 Introduction

Deep Learning (DL) has transformed medical imaging due its power of identifying patterns in the image that are not immediately visible to the human eye, resulting in remarkable success for segmentation, disease detection and diagnosis [16]. However, medical datasets are inherently heterogeneous, influenced by factors such as scanners, acquisition protocols, and patient demographics. These factors introduce spurious correlations, leading DL models to rely on superficial patterns, known as shortcuts, instead of true causal relationships [11,17,19]. Consequently, models that perform well within a training domain often fail to generalize to new distributions [5,13]. For instance, a model trained on a cohort where a disease is more prevalent in male patients may associate the disease with male sex, yielding to biased predictions when applied to mixed-sex populations. In this scenario, instead of learning the true complex anatomical features indicative of the disease, the model exploits a demographic shortcut that leads to biased decision-making in patient care. In response, causal representation learning aims to address this issue by disentangling causal relationships from spurious correlations to improve robustness under distribution shifts and counterfactual changes. Existing strategies include counterfactual augmentation [20,21], dataset rebalancing [1,14], adversarial training [8], and dependence minimization [7,12].

The previous Mutual Information Minimization Model (MIMM) [7] addressed shortcut learning by disentangling a primary task from a *single* spurious factor by minimizing mutual information (MI). Yet, real-world medical datasets often contain multiple, entangled spurious correlations. To this end, we propose MIMM-X, a framework that simultaneously disentangles and observes a primary task from multiple spuriously correlated factors with minimal computational overhead. We validate MIMM-X across three large-scale medical imaging datasets using two modalities (brain MRI from the German National Cohort (NAKO) [6] and UK Biobank (UKB) [15], and chest X-ray from CheXpert [10]). We demonstrate that MIMM-X mitigates shortcut learning of the primary task and improves generalization across distribution shifts caused by (i) induced spurious correlations and (ii) factors that may naturally act as spuriously correlated factors, without modifying their distribution in our experimental setup.

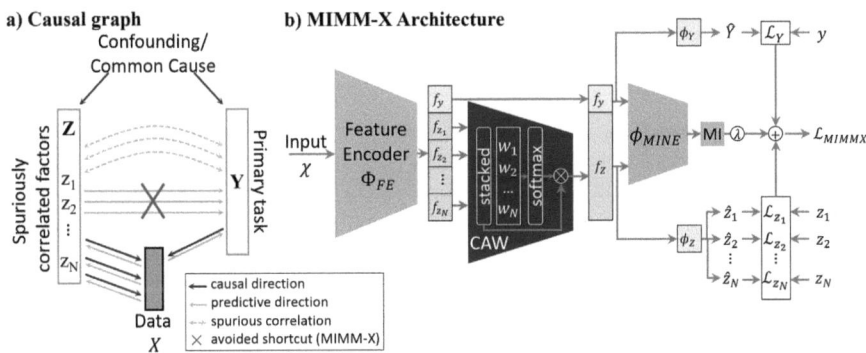

Fig. 1. (a) Causal graph with multiple spurious correlations present. The aim is to avoid predictions based on shortcuts introduced by spurious correlations. (b) Our MIMM-X model promotes causal feature learning by minimizing the MI between the desired primary task y and N spurious correlations Z.

2 Methods

We propose MIMM-X, designed to mitigate multiple spurious correlations $(z_i)_{i=1}^N$ while learning causal representations for a primary task y from a vision input X (Fig. 1a). It is an extension of MIMM [7], which, in comparison, can only handle a single known spurious correlation. As shown in Fig. 1b, MIMM-X consists of a feature encoder ϕ_{FE}, two classification heads ϕ_Y and ϕ_Z, a mutual information estimator ϕ_{MINE}, and a Confounder Attention Weighter module (CAW).

2.1 Architectural Components

Feature Encoder $\phi_{FE}(X, \theta_{FE})$ maps an image X to a feature vector f, partitioned into $N+1$ equal-sized subvectors. The subvector f_y encodes the primary task, while $f_Z = [f_{z_1}, \ldots, f_{z_N}]^T$ represents the N spuriously correlated factors.

Classification Heads. The primary task y is predicted from f_y through the classification head ϕ_Y with a log-softmax output. All N spuriously correlated factors are predicted from the batch-normalized f_Z with a linear multi-task classification head ϕ_Z and log-softmax.

Confounder Attention Weighter (CAW). Before passing f_z to the classification head ϕ_z, CAW assigns adaptive attention weights to each spuriously correlated factor. A learnable parameter vector of size N is normalized via softmax to produce attention weights $\mathbf{w} \in \mathbb{R}^N$. Each spuriously correlated feature subvector f_{z_i} is then weighted by its corresponding attention weight w_i to emphasize more relevant spurious factors. The attention weights are learned jointly with ϕ_{FE}.

Mutual Information Estimation ϕ_{MINE}. To enforce independence between f_y and f_Z, we minimize their mutual information (MI). We estimate a single MI value between f_y and the stacked f_Z using a MINE model [2] to keep computational efficiency while scaling to multiple spurious correlations.

2.2 Training Process

MIMM-X uses alternating updates: $\phi_{FE}, \phi_Y, \phi_Z$ and CAW are updated on one batch, followed by $N_B - 1$ updates of ϕ_{MINE} where $N_B \in \mathbb{N}$ denotes the number of batches per iteration cycle. The overall loss $\mathcal{L}_{\text{MIMM-X}}$ combines cross-entropy terms $\mathcal{L}_0$ for y and $\mathcal{L}_i$ for $i = 1, \ldots, N$ with MI penalty weighted by λ:

$$\mathcal{L}_{\text{MIMM-X}} = \gamma_0 \mathcal{L}_0 + \sum_{i=1}^{N} \gamma_i \mathcal{L}_i + \lambda \phi_{MINE}(f_y, f_Z). \tag{1}$$

Inspired by GradNorm [4], task-specific dynamic loss scaling (DLS) adjusts each task's contribution. Scaling factors γ_i are defined as:

$$\gamma_i = \left(\frac{\mathcal{L}_i}{\overline{\mathcal{L}}}\right)^{\alpha_i}, \quad \text{where } \overline{\mathcal{L}} = \frac{1}{N+1} \sum_{i=0}^{N} \mathcal{L}_i, \quad \text{and } \alpha_i = \begin{cases} \alpha_Y, & \text{if } i = 0, \\ \alpha_Z, & \text{if } i = 1, \ldots, N \end{cases} \tag{2}$$

with dynamic α_Y for the primary task:

$$\alpha_Y = \alpha_{Y,\text{initial}} + \left(\frac{\text{epoch}}{N_{\text{epoch}}} \cdot \beta_Y\right), \tag{3}$$

where β_Y controls the increasing emphasis on y over training epochs.

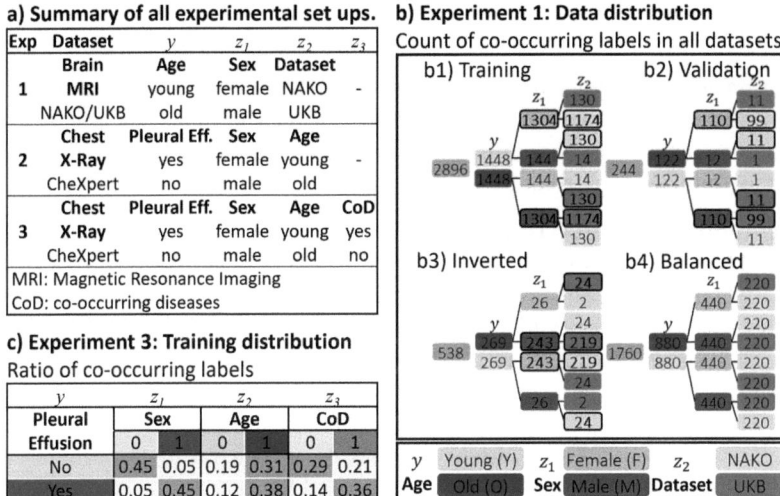

Fig. 2. (a) Summary of experiments, primary tasks y and spurious correlations $(z_i)_{i=1}^N$. (b) Experiment 1: Data distributions (absolute sample counts). (c) Experiment 3: Training data composition with synthetic and natural correlations.

3 Materials

We evaluated MIMM-X in three experiments with two modalities (Fig. 2a).

Training Setup. In Experiments 1 2, we introduced two strong correlations between y and (z_1, z_2) by sub-sampling training data. For each class of y, 90% of samples were drawn from a specific class of z_1 and z_2, creating spurious correlations between y and z_1, z_2 (see Fig. 2b1). Experiment 3 is based on one synthetic spurious correlation and additionally controls for two potentially naturally occurring spurious correlations (Fig. 2c). All experiments were run on a single NVIDIA GeForce RTX 3090 GPU.

Evaluation. We evaluate all models on three types of datasets: *Validation set (Val.)*: same spurious correlations as training (Fig. 2b2); *Inverted set (Inv.)*: inverse correlations (Fig. 2b3); *Balanced set (Bal.)*: no spurious correlation (Fig. 2b4). If no performance drop is experienced when shifting from the validation distribution to the inverted test distribution, the model successfully avoided learning the shortcut introduced by the spurious correlation.

Evaluation Against Comparison Methods. We compare MIMM-X to four methods. *Baseline* uses the same architecture without the MI penalty. *Rebalancing* balances the training set through resampling of under-represented spuriously correlated factor classes. *Distance correlation (dCor)* [12,18] replaces MI with dCor as the dependence penalty. *MIMM* [7] minimizes MI between the primary task and a single spurious factor; we trained separate models for z_1 and z_2, reporting the mean performance for y and individual results for each z_i.

Evaluation of Disentanglements. We assess cross-predictive performance: using f_y to predict $(z_i)_{i=1}^N$, and $f_{(z_i)_{i=1}^N}$ to predict y on the balanced test set. Ideal disentanglement yields random-guess performance, meaning that no information about the opposed task remains in the specific feature subvector.

Experiment 1 (Brain MRI). We used central 2D axial slices from 3D brain MRI (NAKO/UKB), resized to 256 × 256 and z-score normalized. As primary task y, we predicted age as binary group (young: <51 years/old: >57 years) from the extracted slice, while we chose z_1: sex (female/male) and z_2: dataset (NAKO/UKB) as spuriously correlated factors. We used 2,896 training, 244 validation, 538 inverted, and 1,760 balanced samples (Fig. 2a). We applied the same feature encoder as in [7], which is based on four convolutional layers. Training was performed using a batch size of 150 and $N_B = 6$. DLS hyperparameters are set to $\alpha_Y = 0.3$, $\alpha_Z = 0.8$, $\beta_Y = 0.01$, and $\lambda = 1.5$.

Experiment 2 (Chest X-Ray). We used chest X-rays from CheXpert [10] downsampled to 96 × 96 to predict pleural effusion (yes/no) as primary task y. The training set was strongly correlated by z_1: sex (female/male) and z_2: age (young: <50 years/old: > 60 years).

Our training set included 4,126 X-rays, the validation 408 X-rays, the inverted test set of 2,692 X-rays, and the balanced set 4,432 X-rays. The feature encoder was a DenseNet-121 [9], which was trained for 300 epochs with a learning rate of 10^{-5} and a batch size of 100 using $N_B = 5$. The DLS hyperparameter were $\alpha_Y = 0.3$, $\alpha_Z = 0.7$, $\beta_Y = 0.1$, and $\lambda = 1.5$.

Experiment 3 (Chest X-Ray). We extended Experiment 2 to simulate real-world complexity. We kept pleural effusion as primary task y and introduced a single spurious correlation with sex (z_1) in the training set. Additionally, we controlled for two further factors that are often spuriously correlated with disease labels in clinical datasets: age (z_2) and presence of any of the remaining 12 co-occurring lung diseases (CoD, z_3) listed in [3]. These factors were not explicitly correlated with y. Their natural distributions were left unchanged to preserve clinically realistic associations. Importantly, z_2 and z_3 may still act as latent spurious correlates, potentially biasing predictions. This setup enables us to evaluate whether MIMM-X can disentangle y not only from the known spurious factor z_1, but also from naturally co-occurring variables that are not manually manipulated. The training distribution is shown in Fig. 2c. The dataset included 9,850 training, 1,114 validation, 4,064 inverted, and 2,204 balanced test samples.

4 Results and Discussion

Experiment 1. *Evaluation against Comparison Methods* (Table 1a). In this experiment, we aim to predict age group y from brain MRI based on causal features, while avoiding shortcuts through induced spurious correlations: sex

Table 1. Experiment 1 (Brain MRI): Primary task y: age (young/old); spuriously correlated factors z_1: sex (female/male), z_2: dataset (NAKO/UKB).

(a) **Classification accuracy [%]** across different evaluation sets (Val./Inv./Bal.).

Method	DLS	CAW	$f_y \to y$			$f_z \to z_1$			$f_z \to z_2$		
			Val.	Inv.	Bal.	Val.	Inv.	Bal.	Val.	Inv.	Bal.
Baseline	–	–	96.3	27.2	70.2	99.0	96.4	95.9	100.0	100.0	99.9
Baseline	✓	✓	89.6	42.9	66.6	94.2	91.1	90.5	100.0	100.0	99.8
Rebalance	✓	✓	91.8	81.9	87.3	97.5	96.7	97.3	100.0	100.0	100.0
dCor	–	–	96.2	41.7	73.4	98.8	95.7	96.8	100.0	100.0	99.7
MIMM	–	–	74.4	45.7	58.9	88.3	83.8	85.9	100.0	99.6	99.9
MIMM-X	–	–	46.3	54.0	53.5	89.2	88.3	71.6	99.2	98.3	95.8
MIMM-X	–	✓	94.2	33.2	69.0	63.3	57.5	66.5	54.6	44.5	57.3
MIMM-X	✓	–	85.6	74.2	79.8	96.8	96.3	89.4	98.2	98.4	95.4
MIMM-X	✓	✓	84.2	82.8	82.6	93.4	94.6	93.6	100	99.8	99.5

(b) **Disentanglement performance.** Accuracy in [%]. Ideally, near random (50%).

	Baseline			Rebalance			dCor			MIMM			MIMM-X		
	y	z_1	z_2	y	z_1	z_2	y	z_1	z_2	y	z_1	z_2	y	z_1	z_2
f_y	–	70.3	70.0	–	47.6	48.1	–	70.1	67.4	–	40.4	41.4	–	51.8	50.0
f_{z_1}	51.7	–	51.7	49.6	–	49.4	52.3	–	52.4	54.1	–	–	49.6	–	51.5
f_{z_2}	50.0	50.1	–	50.1	50.0	–	50.1	50.0	–	50.7	–	–	49.8	50.1	–

(z_1) and dataset origin, NAKO or UKB (z_2). We first evaluated the impact of our DLS function (Eq. 2) and CAW module within MIMM-X. Although using CAW without DLS yielded the highest validation accuracy, it dropped by over -60% on the inverted distribution, indicating reliance on spurious correlations (z_1, z_2) that no longer hold in this setting. We found that DLS dynamically balances the loss contributions, preventing the primary task signal from being dominated by easier-to-learn spurious correlations with z_1 and z_2.

For the comparison methods, Baseline and dCor, performance drops by -61.6% and -54.5% from validation to inverted dataset. The previous model MIMM, designed for a single spurious correlation, also performs poorly in this multi-correlated setting, with performance of y below random guessing on the inverted set. Rebalancing is more robust (-9.9% drop). However, MIMM-X with CAW and DLS effectively prevents shortcut learning and achieves the best generalization, with only a 1.4% performance drop across distributions and the highest accuracy (82.8%) on the inverted set. As expected, the prediction performance of the spuriously correlated factors remains high for all methods and datasets, as these tasks are inherently less complex to learn than age.

Evaluation Disentanglement (Table 1b). We evaluated disentanglement via cross-prediction: predicting z_1/z_2 from f_y and y from f_{z_1}/f_{z_2} to assess residual shortcut information. Only MIMM-X and Rebalancing achieved near random-guess performance when predicting z_1/z_2 from f_y, confirming effective disentanglement of causal and spurious features. However, compared to MIMM-X, Rebalancing increases the training sample size by 224.3%, requiring more compute and training time. In contrast, Baseline and dCor retained substantial shortcut information. The t-SNE plots (Fig. 3) show that MIMM-X successfully disentangled f_y from sex (z_1) and dataset information (z_2), as no clear separation is visible when coloring f_y by the classes of z_1 (female/male) and z_2 (NAKO/UKB). Other reference methods still allow a separation based on spuriously correlated factors. These results highlight the effectiveness of MIMM-X in mitigating shortcut learning by better disentangling primary features of the age group from spurious correlations of sex and dataset.

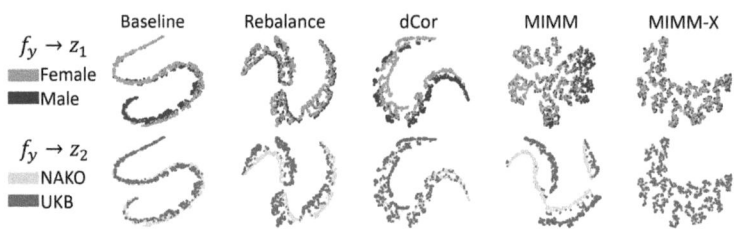

Fig. 3. Experiment 1: t-SNE of f_y colored by spurious factors z_1: sex, z_2: dataset. Ideally, f_y is independent of z_1/z_2, meaning no visual class separation. While reference methods show clusters, MIMM-X is free from spurious information

Experiment 2. *Evaluation against Comparison Methods* (Table 2a).
For our CheXpert experiment with two introduced spurious correlations (z_1:sex, z_2:age), we expected that shortcut learning is prevented if the performance on inverted and balanced test set for pleural effusion y remained high. Similar to Experiment 1, we found that all comparison methods except for Rebalancing are near 50% on the inverted distribution. Only Rebalancing and MIMM-X generalized well, with MIMM-X achieving the highest inverted accuracy (70.4%).

Evaluation Disentanglement (Table 2b). To evaluate disentanglement, we again predicted y, z_1, and z_2 from the opposed subvectors, i.e. $f_y \to z_1/z_2$; $f_{z_1} \to y/z_2$; $f_{z_2} \to y/z_1$. Disentanglement results confirmed that MIMM-X and Rebalancing were closest to random guessing across all tasks, while other methods leave substantial spurious information in f_y (accuracies ¿60%).

Experiment 3. *Evaluation against Comparison Methods* (Table 3a). In this experiment, we investigated a more realistic setting, where we only induced one spurious correlation, z_1: sex, while z_2: age and z_3: CoD retained as naturally occurring correlations, i.e., their distribution in the training set was left unchanged (Fig. 2c). We found that MIMM-X achieved highest performance on the inverted set with 70.2% for the prediction of pleural effusion. Only Rebalancing achieved a similar performance, however, its performance on z_1 and z_2

Table 2. Experiment 2 (X-ray): Primary task y: pleural effusion; spuriously correlated factors z_1: sex (female/male), z_2: age (young/old)

(a) Classification accuracy [%] across different evaluation sets (Val./Inv./Bal.).

Method	$f_y \to y$ Val. Inv. Bal.	$f_z \to z_1$ Val. Inv. Bal.	$f_z \to z_2$ Val. Inv.	Bal.
Baseline	82.6 42.6 64.0	79.7 74.8 73.6	74.4 70.5	57.4
Rebalance	77.7 70.3 76.4	70.8 69.3 71.2	76.5 71.9	74.8
dCor	87.8 42.2 69.0	80.0 62.3 65.1	83.7 63.8	64.5
MIMM	87.2 51.0 71.4	72.1 68.9 66.7	63.6 64.1	53.1
MIMM-X (ours)	86.3 70.4 76.5	79.1 76.2 77.8	78.2 74.8	77.2

(b) Disentanglement performance. Accuracy in [%]. Ideally, near random (50%).

	Baseline y z_1 z_2	Rebalance y z_1 z_2	dCor y z_1 z_2	MIMM y z_1 z_2	MIMM-X y z_1 z_2
f_y	– 64.5 62.5	– 54.9 53.9	– 66.4 64.0	– 61.1 61.0	– 54.8 54.3
f_{z_1}	57.9 – 55.5	48.8 – 49.4	56.8 – 58.7	49.8 – –	50.0 – 50.0
f_{z_2}	52.9 62.6 –	52.7 50.5 –	65.8 65.0 –	50.4 – –	49.6 50.1 –

Table 3. Experiment 3 (X-ray): Primary task y: pleural effusion; spuriously correlated factors z_1: sex (female/male), z_2: age (young/old), z_3: CoD (yes/no)

(a) Classification accuracy [%] across different evaluation sets (Val./Inv./Bal.).

Method	$f_y \to y$ Val. Inv. Bal.	$f_z \to z_1$ Val. Inv. Bal.	$f_z \to z_2$ Val. Inv.	Bal.	$f_z \to z_3$ Val. Inv. Bal.
Baseline	85.4 57.4 71.5	85.7 62.2 58.6	73.2 74.4	73.6	63.5 61.0 62.5
Rebalance	74.4 69.4 71.4	76.8 71.6 74.0	71.7 73.0	72.1	57.7 55.5 57.1
dCor	87.0 54.9 70.9	87.5 72.0 74.0	81.0 80.7	78.0	61.5 57.6 57.5
MIMM-X (ours)	81.7 70.2 72.9	81.5 72.7 74.2	81.4 78.3	79.4	58.9 57.5 58.1

(b) Disentanglement performance. Accuracy in [%]. Ideally, near random (50%).

	Baseline y z_1 z_2 z_3	Rebalance y z_1 z_2 z_3	dCor y z_1 z_2 z_3	MIMM-X y z_1 z_2 z_3
f_y	– 66.6 60.9 60.3	– 45.7 53.5 50.8	– 70.0 62.2 61.3	– 52.6 53.1 50.9
f_{z_1}	65.9 – 62.8 60.6	68.5 – 62.8 58.5	62.6 – 60.8 59.5	64.7 – 59.9 56.0
f_{z_2}	59.9 59.3 – 57.1	57.1 55.0 – 53.7	57.5 53.8 – 52.5	56.9 53.5 – 52.3
f_{z_3}	55.4 53.9 60.1 –	50.1 50.1 56.2 –	50.0 50.0 56.0 –	50.0 50.0 56.2 –

is below MIMM-X. Other methods showed primary task performance below < 60%.

Evaluation Disentanglement (Table 3b). We evaluated disentanglement again by predicting the opposed task of the subvectors for the primary task and the three spuriously correlated factors. We found that f_y of MIMM-X is closest to random guessing compared to the other comparison methods.

We acknowledge some limitations. MIMM-X currently relies on specifying confounding factors in advance. Extending it to discover unknown spurious correlations automatically and to handle more than three, as well as more complex confounding structures, will be explored in future work. Additionally, while dataset rebalancing can achieve comparable performance in some settings, it requires significantly larger training sets, increasing training time and computational cost. More advanced sampling strategies may mitigate this drawback, but they still rely on explicit knowledge of the spurious attributes and demand careful dataset design. In particular, the latter may not always be achievable in strongly non-homogeneous data cohorts. In contrast, MIMM-X offers a scalable solution that mitigates shortcut learning without modifying the data distribution.

5 Conclusion

This work presents MIMM-X, a scalable method for mitigating multiple spurious correlations and avoiding shortcut learning in DL-based medical image analysis. Our experiments demonstrate that MIMM-X improves robustness and feature disentanglement in scenarios with both induced and naturally occurring correlations, contributing toward causal and fair predictions.

Acknowledgments. This project was conducted with data (Application No. NAKO-195 and NAKO-708) from the German National Cohort (NAKO) (www.nako.de). The NAKO is funded by the Federal Ministry of Education and Research (BMBF) [project funding reference no. 01ER1301A/ B/C and 01ER1511D], federal states of Germany and the Helmholtz Association, the participating universities and the institutes of the Leibniz Association. We thank all participants who took part in the NAKO study and the staff of this research initiative.
This work was carried out under U.K. Biobank Application 40040. They also thank all participants who took part in the UKB study and the staff in this research program.

Disclosure of Interests. The authors declare no competing interests.

References

1. Ando, S., Huang, C.Y.: Deep over-sampling framework for classifying imbalanced data. In: Ceci, M., Hollmén, J., Todorovski, L., Vens, C., Džeroski, S. (eds.) ECML PKDD 2017. LNCS (LNAI), vol. 10534, pp. 770–785. Springer, Cham (2017). https://doi.org/10.1007/978-3-319-71249-9_46
2. Belghazi, M.I., et al.: Mutual information neural estimation. In: International Conference on Machine Learning, pp. 531–540. PMLR (2018)
3. Chambon, P., et al.: Chexpert plus: augmenting a large chest x-ray dataset with text radiology reports, patient demographics and additional image formats. arXiv preprint arXiv:2405.19538 (2024)
4. Chen, Z., Badrinarayanan, V., Lee, C.Y., Rabinovich, A.: Gradnorm: gradient normalization for adaptive loss balancing in deep multitask networks. In: International Conference on Machine Learning, pp. 794–803. PMLR (2018)

5. Cohen, J.P., Hashir, M., Brooks, R., Bertrand, H.: On the limits of cross-domain generalization in automated x-ray prediction. In: Medical Imaging with Deep Learning, pp. 136–155. PMLR (2020)
6. geschaeftsstelle@ nationale-kohorte. de, G.N.C.G.C.: The german national cohort: aims, study design and organization. Euro. J. Epidemiol.**29**(5), 371–382 (2014)
7. Fay, L., Cobos, E., Yang, B., Gatidis, S., Küstner, T.: Avoiding shortcut-learning by mutual information minimization in deep learning-based image processing. IEEE Access **11**, 64070–64086 (2023)
8. Ganin, Y., Lempitsky, V.: Unsupervised domain adaptation by backpropagation. In: International Conference on Machine Learning, pp. 1180–1189. PMLR (2015)
9. Huang, G., Liu, Z., Van Der Maaten, L., Weinberger, K.Q.: Densely connected convolutional networks. In: Proceedings of the IEEE Conference on Computer Vision and Pattern Recognition, pp. 4700–4708 (2017)
10. Irvin, J., , et al.: Chexpert: a large chest radiograph dataset with uncertainty labels and expert comparison. In: Proceedings of the AAAI Conference on Artificial Intelligence, vol. 33, pp. 590–597 (2019)
11. Kumar, A., et al.: Debiasing counterfactuals in the presence of spurious correlations. In: Workshop on Clinical Image-Based Procedures, pp. 276–286. Springer (2023). https://doi.org/10.1007/978-3-031-45249-9_27
12. Müller, S., Fay, L., Koch, L.M., Gatidis, S., Küstner, T., Berens, P.: Benchmarking dependence measures to prevent shortcut learning in medical imaging. In: International Workshop on Machine Learning in Medical Imaging, pp. 53–62. Springer (2024). https://doi.org/10.1007/978-3-031-73290-4_6
13. Pooch, E.H.P., Ballester, P., Barros, R.C.: Can we trust deep learning based diagnosis? the impact of domain shift in chest radiograph classification. In: Petersen, J., et al. (eds.) TIA 2020. LNCS, vol. 12502, pp. 74–83. Springer, Cham (2020). https://doi.org/10.1007/978-3-030-62469-9_7
14. Sagawa, S., Koh, P.W., Hashimoto, T.B., Liang, P.: Distributionally robust neural networks for group shifts: On the importance of regularization for worst-case generalization. arXiv preprint arXiv:1911.08731 (2019)
15. Sudlow, C., et al.: Uk biobank: an open access resource for identifying the causes of a wide range of complex diseases of middle and old age. PLOS Med. **12**(3), 1–10 (2015). https://doi.org/10.1371/journal.pmed.1001779
16. Suganyadevi, S., Seethalakshmi, V., Balasamy, K.: A review on deep learning in medical image analysis. Inter. J. Multimedia Inform. Retrieval **11**(1), 19–38 (2022)
17. Sun, S., Koch, L.M., Baumgartner, C.F.: Right for the wrong reason: can interpretable ml techniques detect spurious correlations? In: International Conference on Medical Image Computing and Computer-Assisted Intervention, pp. 425–434. Springer (2023). https://doi.org/10.1007/978-3-031-43895-0_40
18. Székely, G.J., Rizzo, M.L., Bakirov, N.K.: Measuring and testing dependence by correlation of distances. Annals Statist. (2007)
19. Veitch, V., D'Amour, A., Yadlowsky, S., Eisenstein, J.: Counterfactual invariance to spurious correlations: why and how to pass stress tests. arXiv preprint arXiv:2106.00545 (2021)
20. Xu, M., et al.: Adversarial domain adaptation with domain mixup. In: Proceedings of the AAAI Conference on Artificial Intelligence, vol. 34, pp. 6502–6509 (2020)
21. Yao, H., et al.: Improving out-of-distribution robustness via selective augmentation. In: International Conference on Machine Learning, pp. 25407–25437. PMLR (2022)

Predicting Patient Self-reported Race From Skin Histological Images with Deep Learning

Shengjia Chen[1,2], Ruchika Verma[1,2], Kevin Clare[4], Jannes Jegminat[1,2], Eugenia Alleva[1,2], Kuan-lin Huang[3], Brandon Veremis[4], Thomas Fuchs[1,2], and Gabriele Campanella[1,2(✉)]

[1] Windreich Department of Artificial Intelligence and Human Health, Icahn School of Medicine at Mount Sinai, New York, USA
gabriele.campanella@mssm.edu
[2] Hasso Plattner Institute for Digital Health at Mount Sinai, Icahn School of Medicine at Mount Sinai, New York, USA
[3] Mount Sinai Center for Transformative Disease Modeling, Icahn School of Medicine at Mount Sinai, New York, USA
[4] Department of Pathology, Molecular and Cell-Based Medicine, Mount Sinai Health System, New York, USA

Abstract. Artificial Intelligence (AI) has demonstrated success in computational pathology (CPath) for disease detection, biomarker classification, and prognosis prediction. However, its potential to learn unintended demographic biases, particularly those related to social determinants of health, remains understudied. This study investigates whether deep learning models can predict self-reported race from digitized dermatopathology slides and identifies potential morphological shortcuts. Using a multisite dataset with a racially diverse population, we apply an attention-based mechanism to uncover race-associated morphological features. After evaluating three dataset curation strategies to control for confounding factors, the final experiment showed that White and Black demographic groups retained high prediction performance (AUC: 0.799, 0.762), while overall performance dropped to 0.663. Attention analysis revealed the epidermis as a key predictive feature, with significant performance declines when these regions were removed. These findings highlight the need for careful data curation and bias mitigation to ensure equitable AI deployment in pathology. Code available at: https://github.com/sinai-computational-pathology/CPath_SAIF.

Keywords: Computational Pathology · AI Fairness · Dermatopathology

1 Introduction

Bias and disparities in Machine Learning (ML)-based biomedical and healthcare applications have been widely studied by stratifying model performance across demographic groups [24]. Recent studies have shown that ML models can propagate or even exacerbate existing healthcare inequalities due to dataset bias, arising from differences in disease prevalence, clinical presentation, and annotation inconsistencies across demographic groups [16,21]. While algorithmic fairness techniques have been explored to mitigate bias [30], several studies have also demonstrated that featureconfounder correlations, such as the presence of treatment artifacts or institution-specific markers, can undermine model generalizability and fairness [12,14,22].

In computational pathology (CPath), deep learning (DL) models have shown promise in disease detection [13], biomarker classification [7], and prognosis prediction [25], but demographic disparities in performance have also been reported in recent studies [26]. While biases and demographic shortcuts are well-studied in medical imaging [31], particularly radiology [1,8], similar investigations in histopathology remain limited. Histological slides capture complex tissue morphology, cellular structures, and microenvironmental characteristics, but it is unclear whether these reflect demographic variations. Identifying such associations is crucial to understand confounders that may influence differential model performance in CPath [28].

In this study, we investigate whether the DL models can infer self-reported race from histological images using skin histology data collected across multiple sites within a health system, without specific curation. Skin histology provides a unique opportunity for this analysis, as characteristics related to melanin and pigmentationwhile visibly distinct in clinical dermatology [11]are not readily apparent in histological images, making it unclear whether DL models can still capture race-associated patterns. By focusing on a single organ system, we effectively control for potential confounding variables that would present greater challenges in a more heterogeneous dataset. Using widely validated tile-level foundation models (FMs) in CPath [2] combined with explainable attention-based model AB-MIL [5,15], we examine whether tissue and cellular features can predict self-reported race. Furthermore, we implement a histomorphological phenotype learning framework [6] to identify morphologies associated with high attention regions, providing biological insights into model behavior.

Related Work. Recent studies have shown that DL models can predict self-reported race with high accuracy across medical imaging modalities, particularly in radiology [32]. Adleberg et al. [1] reported an AUC of 0.911 for race prediction using chest radiographs, a capability that persists across modalities even when undetectable to human experts [8]. Beyond classification, race-related feature encodings have been observed in chest X-ray foundation models [9] and brain age prediction models trained on MRI, with both showing performance disparities and statistically significant distribution shifts across demographic subgroups [23]. In histopathology, stain variability and site-specific digital signatures can correlate with ethnicity and inflate model performance [14,17]. Additionally, models trained for diagnostic tasks can encode racial information, with diagnostic accuracy positively associated with race prediction performance, even after mitigation efforts [26]. Extending these findings to CPath, our work investigates histomorphological features associated with self-reported race in dermatopathology, aiming to identify potential biological confounders and assess their influence on model predictions.

2 Methods

Dataset. Self-reported race, a social construct with known correlations to differential health outcomes and a widely recognized social determinant of health, was collected from patient records and questionnaires at Mount Sinai health system (MSHS). Patients with self-reported race equal to "unknown" or "not reported" were removed. Our private dataset consists of digitized slides from all available skin specimens, assembled from multiple sites in New York city (NYC), with all scanned on a Philips Ultrafast scanner. The dataset exhibits a diverse racial distribution with the overall patient population at MSHS, closely matching the city's demographics. Although the White group is slightly overrepresented (39.3%), this imbalance is relatively minor compared to other widely used histological datasets, such as TCGA, where 73.7% of samples are from White patients. Self-reported race data are provided for comparison in Table 1.

The dataset was generated from all available dermatopathology specimens within the health system, rather than being curated for a specific disease or prediction task. This includes a wide range of skin conditions such as hemorrhoids, melanoma, basal cell carcinoma (BCC), seborrheic keratosis, squamous cell carcinoma (SCC), and various types of inflammatory and infectious dermatoses. Additionally, the potential site-specific signature, as suggested in [14], has been controlled for since all slides collected from different sites were stained and digitized in a central laboratory within the health system.

Table 1. Summary of the skin dataset by self-reported race compared with Mount Sinai healthcare system, New York city population, and public source (TCGA).

Self-reported Race	Skin Cohort		MSHS	NYC	TCGA
	# Slides (%)	Patients %		Population %	
White	2,151 (40.8%)	39.3	43.1	31.2	73.7
Black	1,015 (19.3%)	19.0	21.7	29.9	10.3
Hispanic/Latino	868 (16.5%)	16.8	18.5	21.0	8.5
Asian	687 (13.1%)	15.7	10.3	5.7	7.1
Other	543 (10.3%)	9.3	6.4	4.5	1.8
Total Number	5,266	2,471	114,947	8M	23,276

Experimental Setup. To investigate the capability of DL models to classify self-reported race from histological images, we implemented a classification pipeline leveraging FM for feature extraction. Each slide was assigned a label corresponding to the self-reported race of the patient. The dataset was split 80/20 for training and validation at the patient level, with no separate test set allocated since generalization was not the focus. Tissue tiles were extracted at 20x magnification, and tile-level embeddings were generated using four pre-trained FMs: SP22M [3], UNI [4], GigaPath [29], and Virchow [27], followed by an attention-based MIL (AB-MIL) model [15] for slide-level aggregation. We selected AB-MIL due to its ability to efficiently learn informative tile-level attention scores that highlight discriminative regions within a slide while maintaining interpretability. The attention mechanism allows the model to quantitatively assess each tiles contribution to the slide-level racial prediction. During training, a weighted loss function was applied to ensure class balance, and models were trained for 40 epochs using the AdamW optimizer [18] with an initial learning rate of 0.0005, a 5-epoch warm-up, and a cosine decay schedule. A batch size of 512 was employed, and the final model checkpoint was evaluated on the validation set for performance and attention analysis. To ensure reproducibility and stability, Xavier initialization [10] was applied with three fixed random seeds (0, 42, 2025), and output probabilities and attention scores were averaged across these runs. All training was conducted on a single H100 GPU.

UMAP Visualization. To better understand the histological patterns that are important for the self-reported race prediction task, we utilized a histomorphological phenotype learning framework [6] to efficiently analyze regions of high attention. This tool also enables the efficient segmentation of tile-level histological structures, allowing us to study the relative attention given to different tissue compartments. Instead of training a segmentation model from scratch, we leveraged SP22M [3] to extract tile features, which were then projected into a 2D UMAP space [20] for visualization. Pathologists annotated a few landmark tiles to identify key tissue structures, allowing us to locate similar tiles in the UMAP space. Through iterative refinement, a Random Forest classifier was trained to segment regions of interest (ROI) based on UMAP-embedded tile features, defining ROIs as areas where at least 20% of pixels corresponded to a given morphological class. Representative morphological classes identified in the UMAP space included epidermis, inflammation, gastrointestinal (GI) tissue, bone, adipose tissue (fat), blood, smooth muscle, skeletal muscle, ducts, and oncocytes, as well as common artifacts such as ink, cautery, and coverslip edges. Two pathologists validated these annotations before proceeding with stratified attention analysis.

Attention Scores and Distribution Analysis. The attention score for each self-reported race class was obtained from AB-MIL [15], incorporating a multi-head mechanism similar to CLAM [19] to output distinct attention scores for each race groups. Each tile within a slide received an attention score corresponding to the race prediction head, indicating its contribution to the model's classification decision. To enable cross-slide comparisons, attention scores were normalized across all tiles in the validation dataset to a [0,1] range. We then compared mean attention scores between ROI and non-ROI areas to assess the relationship between attention and tissue morphology, investigating whether specific tissue types contributed more significantly to the model's decision-making process.

3 Results

To evaluate the potential bias of disease distribution on race prediction, we curated three versions of datasets. Model performance was evaluated using one-vs-rest (OvR) area under the curve (AUC), and results are summarized in Table 2. For each row, the mean area under the curve (AUC) score is reported based on 1000 bootstrap iterations. On average, 9,647 tiles were extracted at 20 magnification per slide (min: 46, median: 8,067, max: 45,172), depending on the size of the main tissue area.

Table 2. Model performance across three dataset curations. AUC is one-vs-all, accuracy is balanced accuracy.

Experiment	Encoder	AUCs by Racial Groups					Overall Metrics	
		White	Black	Hispanic	Asian	Other	AUC	Accuracy
Exp1	SP22M	0.772	0.785	0.586	0.805	0.547	0.699	0.395
	UNI	0.797	0.791	0.607	0.791	0.603	**0.718**	**0.400**
Uncurated	GigaPath	0.801	0.753	0.598	0.801	0.522	0.695	0.388
	Virchow	0.784	0.749	0.591	0.783	0.579	0.697	0.392
	Average	**0.789**	0.770	0.596	**0.795**	0.563	0.702	0.394
Exp2	SP22M	0.744	0.751	0.569	0.701	0.577	0.668	0.368
	UNI	0.760	0.773	0.560	0.715	0.569	**0.676**	**0.380**
Balance	GigaPath	0.734	0.739	0.581	0.753	0.559	0.673	0.372
Disease	Virchow	0.728	0.753	0.529	0.726	0.590	0.665	0.334
	Average	**0.742**	**0.754**	0.560	<u>0.724</u>	0.574	0.671	0.364
Exp3	SP22M	0.788	0.773	0.584	0.481	0.534	0.632	0.287
	UNI	0.819	0.766	0.654	0.556	0.594	0.678	0.296
Strict	GigaPath	0.791	0.766	0.664	0.650	0.431	0.661	**0.333**
ICD code	Virchow	0.796	0.742	0.656	0.592	0.613	**0.680**	0.293
	Average	**0.799**	**0.762**	0.640	<u>0.570</u>	0.543	0.663	0.302

Exp1 (Uncurated) included all available dermatopathology specimens and yielded the highest overall OvR AUC (0.702), with particularly strong performance in the Asian group (AUC = 0.795). This was attributed to a disproportionately high prevalence of hemorrhoid cases (61%) among Asian patients due to site-specific sampling biases (160 out of 312 Asian patients treated at one site). **Exp2 (Balance Disease)** mitigated disease-related confounding by rebalancing hemorrhoid cases and removing gangrene and sun damage-related conditions disproportionately prevalent in Black and White patients but had low overall occurrence (e.g., melanoma, basal cell carcinoma, squamous cell carcinoma, actinic keratosis, and seborrheic keratosis), resulting in 2,032 patients (W 37.5%, B 19.8%, H/L 17.3%, A 15.1%, O 10.2%). This adjustment led to a decline in overall OvR AUC (0.671), with the Asian group experiencing the largest drop (AUC: 0.795 0.724). In **Exp3 (Strict ICD Code)**, we further restricted the dataset to classical dermatopathology cases (ICD-10 code, L: inflammatory skin diseases, C: skin cancers, D: benign skin growths and disorders), fully removing hemorrhoids (ICD-10 K), and reducing dataset to 800 patients (W 46.9%, B 19.9%, H/L 19.6%, A 7.2%, O 6.5%). This further reduced the overall OvR AUC

to 0.663, with the Asian group showing the most pronounced decline (0.570), whereas the White group maintained consistently high performance (0.799).

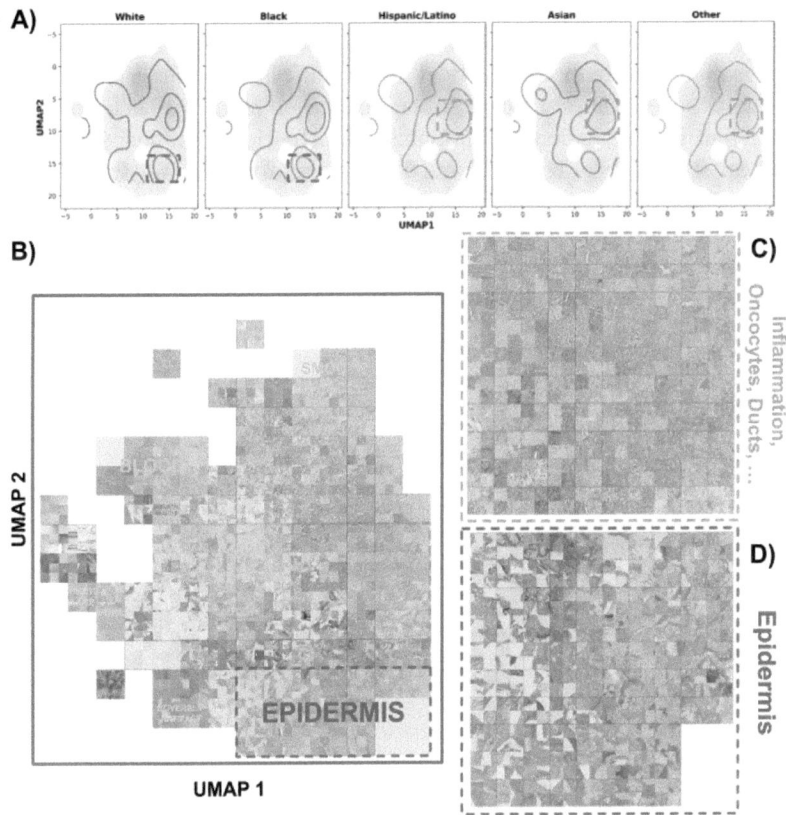

Fig. 1. UMAP visualization of attention scores. (A) Density plot with a grayscale KDE background representing the overall distribution. Contour lines were generated for high-attention tiles within each racial group. (B) Grid plot visualizing representative samples from different UMAP regions. (C, D) Zoomed-in grid plots highlighting regions that received high model attention. GI: gastrointestinal tract.

UMAP Visualization of Attention and Morphological Patterns. Visual inspection of attention scores suggested a spatial association between high attention and tissue morphology across racial groups. To investigate this, we projected attention scores into a lower-dimensional UMAP space using SP22M encoder due to its lightweight architecture (22M parameters) and comparable performance. For UMAP generation, 20 slides per racial group were randomly sampled from **Exp3**, with 10,000 tiles per slide. Figure 1A presents the

UMAP projection, where a grayscale kernel density estimation (KDE) background shows the overall data distribution, and colored contour lines highlight regions with the top 10% of attention scores for each racial group. White and Black groups exhibit more concentrated attention clusters, whereas the Hispanic/Latino, Asian, and Other groups display more dispersed attention distributions, suggesting potential histomorphological differences. To further examine morphological differences in model attention, Fig. 1B visualizes pathologist-annotated tissue types (epidermis, inflammation, blood, fat, background, etc.) from different UMAP regions, reinforcing that attention is influenced by distinct histological structures. Figure 1C and D zoom into two high-attention regions identified in the KDE plot. Figure 1C includes diverse histomorphological types (oncocytes, ducts, inflammation), but high attention in this region lacks a clear structural association. In contrast, Fig. 1D corresponds specifically to epidermis, aligning with the strong epidermal attention observed in the White and Black groups.

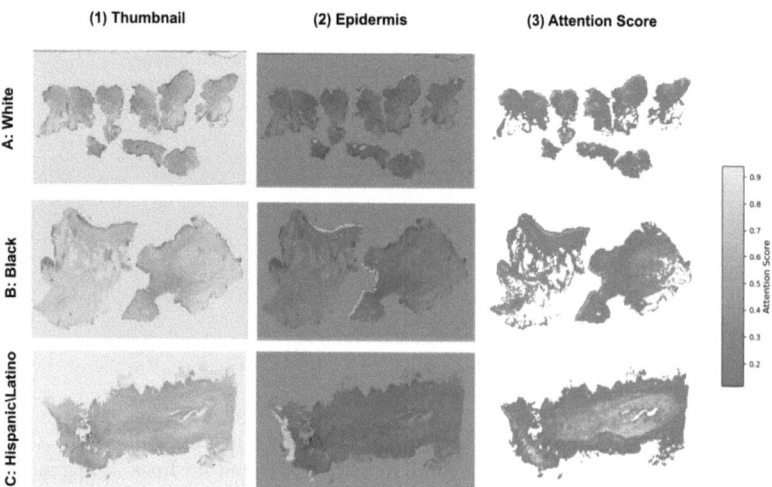

Fig. 2. Whole-slide attention maps for selected examples from **Exp3**. Rows correspond to three selected slides, with columns showing: (1) Whole-slide thumbnail, (2) Binary mask of epidermis detection, and (3) Attention score from a specific racial group. (A) Attention to White, (B) Attention to Black, (C) Attention to Hispanic/Latino.

Attention Distribution Analysis. Figure 2 presents whole-slide attention maps from examples in **Exp3**. In (A), attention to White maps strongly to the epidermis, whereas in (B), attention to Black highlights the epidermis but also extends to other regions. In (C), attention to Hispanic is predominantly observed in non-epidermis regions. To compare attention distribution, we performed a one-sided paired t-test to assess whether epidermal regions received higher attention. Figure 3A presents the median attention score per slide including only slides with more than 15 epidermis tiles were to reduce noise. Across all three experiments, epidermal regions consistently received higher attention than non-epidermis regions, but the magnitude of this difference decreased from **Exp1** to **Exp3**. In **Exp3**, the effect was significant only for the White and Black groups, while the Asian group exhibited lower attention in epidermis than non-epidermis regions. Figure 3B further highlights the importance of epidermis in self-reported race prediction. When epidermis tiles were completely removed in validation data, model performance dropped by approximately 0.05 across all racial groups and experiments. When only epidermis tiles were retainedwith 85% of slides in validation set containing less than 20% epidermis tilesthe model maintained comparable or even improved performance in some racial groups.

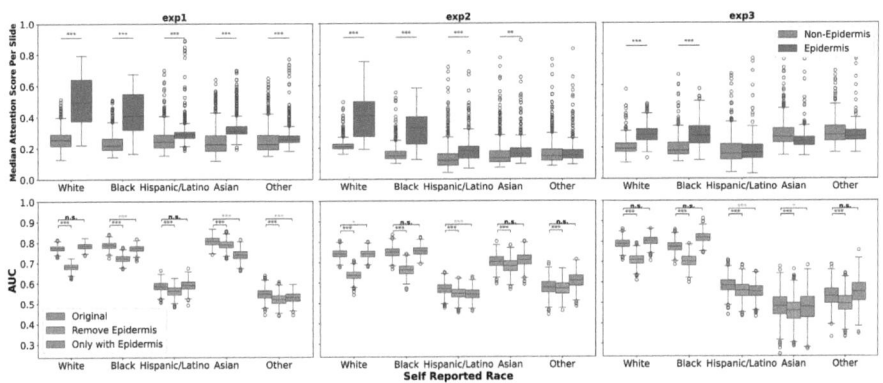

Fig. 3. Attention and ablation analysis across racial groups and experiments. (A) Boxplots comparing the median attention score per slide between epidermis and non-epidermis regions (B) AUCs with epidermis tiles removed (orange), kept only (blue), and compared to original model (green). One-sided paired t-test significance: *: p<0.05, **: p<0.01, ***: p<0.0001, n.s./unlabeled if not significant. (Color figure online)

4 Discussion

In this study, we examined whether DL models can predict self-reported race from digitized dermatopathology slides, independent of pathology task. Unlike previous studies on reporting the differential performance of task-specific models, we explored whether biological correlates of race could be identified in histology images. This is important because task-specific models could exploit these features as shortcuts, leading to unintended biases and disparities in clinical predictions. Our results (**Exp3**) show that self-reported race can be predicted with moderate accuracy (AUC = 0.7), particularly for White (0.80) and Black (0.76) patients. We identified epidermis, which typically constitute 10%20% of tissue in skin histology slides, as the strongest predictive histological component, consistent with the role of melanocytes in skin tone [28]. Across all experiments, White and Black groups consistently had higher prediction performance than Hispanic/Latino and Asian groups, possibly due to more distinct epidermal features in White and Black patients, whereas Hispanic and Asian groups exhibit greater morphological variation, making classification more challenging. Confounding variables in patient sampling and disease presentation also inflated prediction performance. In **Exp1**, the overrepresentation of hemorrhoid cases in Asian patients (61%) likely caused the model to associate race with disease prevalence rather than intrinsic histological differences. **Exp2** rebalanced disease distribution, resulting in a performance drop, suggesting that race labels acted as unintended shortcuts for disease classification. **Exp3**, which applied ICD-10 coding to focus on skin disease cases, was chosen for further investigation.

Our study raises key concerns about demographic shortcuts in CPath but also has several limitations. Self-reported race is a socioeconomic determinant that while identifying potential confounders, also introduces noise. The heterogeneity of the Hispanic/Latino group further complicates isolating specific biological or morphological patterns. Integrating genetic ancestry data alongside self-reported race may provide a more comprehensive understanding of demographic influences in histological analysis. This study focused on skin histology for better control over confounders, but future work should extend to other specimen/organs to determine whether race-associated patterns emerge in other tissues or if skin remains unique due to its link to pigmentation. Additionally, ICD-10 coding has inherent limitations, as it reflects clinical suspicion rather than definitive histological diagnosis.

While histological images may not encode demographic signals as strongly as radiological images [1,8], DL models can still predict self-reported race, likely by leveraging morphological shortcuts such as epidermal structures in skin slides. These findings highlight the need to consider demographic biases in CPath models and the impact of dataset curation on model fairness. Developing bias mitigation methods to address model reliance on demographic shortcuts is crucial for ensuring fairness in CPath applications, and we encourage researchers to carefully account for disease distribution and remain mindful of how AI models may inadvertently learn and exploit sensitive demographic information rather than focusing on disease-related histological features.

Acknowledgments. This work is supported in part through the use of research platform AI-Ready Mount Sinai (AIR.MS) and the expertise provided by the team at the Hasso Plattner Institute for Digital Health at Mount Sinai (HPI.MS). This work was supported in part through the computational and data resources and staff expertise provided by Scientific Computing and Data at the Icahn School of Medicine at Mount Sinai and supported by the Clinical and Translational Science Awards (CTSA) grant UL1TR004419 from the National Center for Advancing Translational Sciences.

Disclosure of Interests. The authors have no competing interests to declare that are relevant to the content of this article.

References

1. Adleberg, J., et al.: Predicting patient demographics from chest radiographs with deep learning. J. Am. Coll. Radiol. **19**(10), 1151–1161 (2022)
2. Campanella, G., et al.: A clinical benchmark of public self-supervised pathology foundation models. arXiv preprint arXiv:2407.06508 (2024)
3. Campanella, G., et al.: Computational pathology at health system scale–self-supervised foundation models from three billion images. arXiv preprint arXiv:2310.07033 (2023)
4. Chen, R.J., et al.: Towards a general-purpose foundation model for computational pathology. Nat. Med. **30**(3), 850–862 (2024)
5. Chen, S., et al.: Benchmarking embedding aggregation methods in computational pathology: A clinical data perspective. arXiv preprint arXiv:2407.07841 (2024)
6. Claudio Quiros, A., et al.: Mapping the landscape of histomorphological cancer phenotypes using self-supervised learning on unannotated pathology slides. Nat. Commun. **15**(1), 4596 (2024)
7. Nahhas, O.S., et al.: From whole-slide image to biomarker prediction: end-to-end weakly supervised deep learning in computational pathology. Nat. Protoc. **20**(1), 293–316 (2025)
8. Gichoya, J.W., et al.: Ai recognition of patient race in medical imaging: a modelling study. Lancet Digital Health **4**(6), e406–e414 (2022)
9. Glocker, B., Jones, C., Roschewitz, M., Winzeck, S.: Risk of bias in chest radiography deep learning foundation models. Radiology: Artifi. Intell. **5**(6), e230060 (2023)
10. Glorot, X., Bengio, Y.: Understanding the difficulty of training deep feedforward neural networks. In: Proceedings of the Thirteenth International Conference on Artificial Intelligence and Statistics, pp. 249–256. JMLR Workshop and Conference Proceedings (2010)
11. Harvey, V.M., et al.: Integrating skin color assessments into clinical practice and research: a review of current approaches. J. Am. Acad. Dermatol. (2024)
12. Hill, B.G., Koback, F.L., Schilling, P.L.: The risk of shortcutting in deep learning algorithms for medical imaging research. Sci. Rep. **14**(1), 29224 (2024)
13. Hosseini, M.S., et al.: Computational pathology: a survey review and the way forward. J. Pathol. Inform. 100357 (2024)
14. Howard, F.M., et al.: The impact of site-specific digital histology signatures on deep learning model accuracy and bias. Nat. Commun. **12**(1), 4423 (2021)
15. Ilse, M., Tomczak, J., Welling, M.: Attention-based deep multiple instance learning. In: International Conference on Machine Learning, pp. 2127–2136. PMLR (2018)
16. Jones, C., Castro, D.C., Sousa Ribeiro, F., Oktay, O., McCradden, M., Glocker, B.: A causal perspective on dataset bias in machine learning for medical imaging. Nat. Mach. Intell. **6**(2), 138–146 (2024)
17. Kheiri, F., Rahnamayan, S., Makrehchi, M., Bidgoli, A.: Bias in histopathology datasets: a comprehensive investigation on possible factors (2024)
18. Loshchilov, I., Hutter, F., et al.: Fixing weight decay regularization in adam. arXiv preprint arXiv:1711.05101 **5** (2017)
19. Lu, M.Y., Williamson, D.F., Chen, T.Y., Chen, R.J., Barbieri, M., Mahmood, F.: Data-efficient and weakly supervised computational pathology on whole-slide images. Nat. Biomed. Eng. **5**(6), 555–570 (2021)
20. McInnes, L., Healy, J., Melville, J.: Umap: Uniform manifold approximation and projection for dimension reduction. arXiv preprint arXiv:1802.03426 (2018)

21. Nazer, L.H., et al.: Bias in artificial intelligence algorithms and recommendations for mitigation. PLOS Digital Health **2**(6), e0000278 (2023)
22. Oakden-Rayner, L., Dunnmon, J., Carneiro, G., Ré, C.: Hidden stratification causes clinically meaningful failures in machine learning for medical imaging. In: Proceedings of the ACM Conference on Health, Inference, and Learning, pp. 151–159 (2020)
23. Piçarra, C., Glocker, B.: Analysing race and sex bias in brain age prediction. In: Workshop on Clinical Image-Based Procedures, pp. 194–204. Springer (2023). https://doi.org/10.1007/978-3-031-45249-9_19
24. Seyyed-Kalantari, L., Zhang, H., McDermott, M.B., Chen, I.Y., Ghassemi, M.: Underdiagnosis bias of artificial intelligence algorithms applied to chest radiographs in under-served patient populations. Nat. Med. **27**(12), 2176–2182 (2021)
25. Song, A.H., et al.: Artificial intelligence for digital and computational pathology. Nat. Rev. Bioeng. **1**(12), 930–949 (2023)
26. Vaidya, A., et al.: Demographic bias in misdiagnosis by computational pathology models. Nat. Med. **30**(4), 1174–1190 (2024)
27. Vorontsov, E., et al.: A foundation model for clinical-grade computational pathology and rare cancers detection. Nat. Med., 1–12 (2024)
28. Williams, K.A., Wondimu, B., Ajayi, A.M., Sokumbi, O.: Skin of color in dermatopathology: does color matter? Hum. Pathol. **140**, 240–266 (2023)
29. Xu, H., et al.: A whole-slide foundation model for digital pathology from real-world data. Nature, 1–8 (2024)
30. Yang, J., Soltan, A.A., Eyre, D.W., Yang, Y., Clifton, D.A.: An adversarial training framework for mitigating algorithmic biases in clinical machine learning. NPJ Digital Med. **6**(1), 55 (2023)
31. Yang, Y., Zhang, H., Gichoya, J.W., Katabi, D., Ghassemi, M.: The limits of fair medical imaging ai in real-world generalization. Nat. Med. **30**(10), 2838–2848 (2024)
32. Zou, J., Gichoya, J.W., Ho, D.E., Obermeyer, Z.: Implications of predicting race variables from medical images. Science **381**(6654), 149–150 (2023)

Robustness and Sex Differences in Skin Cancer Detection: Logistic Regression vs CNNs

Nikolette Pedersen[1], Regitze Sydendal[1(✉)], Andreas Wulff[1,3,4], Ralf Raumanns[1], Eike Petersen[2], and Veronika Cheplygina[1]

[1] IT University of Copenhagen, Copenhagen, Denmark
{nizp,resy,lawu,ralr,vech}@itu.dk
[2] Fraunhofer Institute for Digital Medicine MEVIS, Bremen, Germany
eike.petersen@mevis.fraunhofer.de
[3] Fontys University of Applied Science, Eindhoven, The Netherlands
[4] Eindhoven University of Technology, Eindhoven, The Netherlands

Abstract. Deep learning has been reported to achieve high performances in the detection of skin cancer, yet many challenges regarding the reproducibility of results and biases remain. This study is a replication (different data, same analysis) of a previous study on Alzheimer's disease detection, which studied the robustness of logistic regression (LR) and convolutional neural networks (CNN) across patient sexes. We explore sex bias in skin cancer detection, using the PAD-UFES-20 dataset with LR trained on handcrafted features reflecting dermatological guidelines (ABCDE and the 7-point checklist), and a pre-trained ResNet-50 model. We evaluate these models in alignment with the replicated study: across multiple training datasets with varied sex composition to determine their robustness. Our results show that both the LR and the CNN were robust to the sex distribution, but the results also revealed that the CNN had a significantly higher accuracy (ACC) and area under the receiver operating characteristics (AUROC) for male patients compared to female patients. The data and relevant scripts to reproduce our results are publicly available (https://github.com/nikodice4/Skin-cancer-detection-sex-bias).

1 Introduction

Classification of skin lesions using machine learning (ML), especially deep learning (DL), has shown promising results [6,22]. However, such models have also been shown to suffer from spurious correlations or shortcut learning [4,14,27] as well as biases across patient sex [2,17] or skin types [10].

Bias in medical imaging is an active topic of investigation. Especially in the context of DL, it is challenging to discern what the classifier is learning, rendering bias investigations challenging. For example, Gichoya et al. [12] show that DL models can reliably predict sensitive demographic variables from chest x-rays, even when these variables are not recognizable to clinical experts. While

DL models are highly popular, prior evidence suggests that at least in some applications, simpler models may be more robust to shortcuts and biases while obtaining similar performance [21].

We investigate sex-related performance and robustness in skin lesion classification with two models: (i) a baseline LR with handcrafted features such as asymmetry, border, and colour, and (ii) a CNN trained on raw images of skin lesions. We aim to replicate (*not* reproduce, since we use different data) a previous study on MRI-based Alzheimer's classification [21] which examined the robustness of LR vs CNNs, and found that LR is more robust to different dataset compositions, while CNNs (surprisingly) generally improved their performance for both male and female patients when including more female patients in the training dataset. Our experiments on skin lesion classification with the PAD-UFES-20 dataset [20] reach similar conclusions. Our contributions are as follows:

1. We replicate the results of a robustness study on Alzheimer's classification across sexes in a different medical domain.
2. We highlight previously unreported errors (such as identical lesion IDs with different patients) in the PAD-UFES-20 dataset, demonstrating the importance of data exploration before training.
3. We show that traditional methods such as handcrafted features with LR are still worth examining in the context of skin lesion classification, among others due to their robustness to biases and interpretability.

2 Related Work

There are various dermatological guidelines for diagnosing lesions, including the ABCDE method and the 7-point checklist [1,11]. Early work on automatic diagnosis of skin lesions often focused on implementing "handcrafted" features to reflect these guidelines and applying ML techniques such as random forests. Later, DL became the dominant approach for skin cancer detection, often with CNNs, and more recently, transformers. Promising results have been reported, for example, on the International Skin Imaging Collaboration (ISIC) 2019 data where Pham et al. [22] report to have achieved performances surpassing dermatologists.

However, many methods have been shown to suffer from biases and shortcuts. In skin lesion classification, biases have been shown for age, sex, and Fitzpatrick skin types [2,13], while shortcuts can be introduced by surgical markers [27] or other factors such as the type of dermatoscope used, lightning conditions, the presence of hair, skin tone and others. A categorisation of shortcuts and their influence on skin lesion classification is provided by Jiménez-Sánchez et al. [14].

Findings on sex bias in skin lesion classification diverge, possibly due to differences in the datasets used. Some works [2] show performance disparities, while others no significant differences between male and female patients [23,25]. However, Sies et al. [25] highlight that sex bias cannot be ruled out, especially when considering CNNs.

Several studies highlight the problems with publicly available datasets used for ML, such as underrepresentation of demographic groups or the absence of metadata needed to investigate such biases [9,26]. A popular source of data is the large ISIC archive, though different studies use different subsets of it. However, it contains duplicates, and metadata identifying these is missing [7]. By contrast, the PAD-UFES-20 dataset [20] with over 2K images has a more diverse representation and rich metadata. Another recent and diverse dataset is DDI [10] which, however, contains less than 1000 images.

Table 1. Summary table of data characteristics

Lesion type	No. female patients	No. male patients	Total (%)
Non-cancerous lesions	198	148	346 (29.4)
Cancerous lesions	401	432	833 (70.6)
Total	**599**	**580**	**1179 (100)**

3 Methods

3.1 Data

We use the PAD-UFES-20 dataset [20], containing 2298 images of 1641 skin lesions from 1373 patients. There are six different diagnoses, which we grouped into non-cancerous vs. skin cancer due to the size of the dataset. We removed all patients with missing entries for the gender variable[1]. Furthermore, we removed any duplicates with the same lesion ID, keeping only the first occurrence. The result is a total of 1179 lesions, a summary of the data characteristics can be seen in Table 1.

During data exploration, we observed several inconsistencies in the dataset. These included different patients with the same lesion ID and different lesion IDs for identical images. The creators of PAD-UFES-20 confirmed to us that these were indeed mistakes, and we updated the lesion IDs accordingly[2].

Non-experts (data science students and the authors) created lesion segmentations using LabelStudio [16]. We made the segmentations available on Zenodo[3]. We upsample the non-cancerous lesions, augmenting them by flipping the images, randomly sharpening, and adding Gaussian blur. We did not perform colour augmentation, as this could affect the "ground truth" diagnostic features of the images.

---

[1] Refer to the discussion section for details on our use of the terms 'sex' and 'gender'.
[2] Refer to our Github repository for details.
[3] https://doi.org/10.5281/zenodo.16535326.

3.2 Feature Extraction

We used feature extraction methods related to the ABC criteria, leaving out diameter and evolution, and the 7-point checklist. The features were originally implemented by students in a data science project course on image analysis. We selected several methods based on the degree of quality of the code and feature explanations, and reproducibility of the code. The initial set of features included:

Asymmetry rotates the lesion mask multiple times by 22.5%, "folds" the lesion mask to measure the overlap at this angle, and averages the resulting scores.

Compactness is defined as $c = \frac{p^2}{4\pi A}$, where p is the perimeter of the lesion border and A is the area of the lesion.

Mean and variance of HSV takes the means and variances of the lesion pixels for each channel.

Dominating HSV clusters lesion pixels on their full HSV values using k-Means, and selects the hue value for the largest cluster.

Colour variance clusters lesion pixels using SLIC superpixel segmentation on their RGB values, and takes the mean RGB values for each segment.

Relative colours is defined by [8]. It clusters pixels with SLIC and takes the relative proportions red (F1) and green (F2) inside the lesions. F11 is the difference in red between the skin and the lesion.

Blue whitish veil is defined by [18] and counts the number of pixels where the RGB value satisfies $R > 60, R - 46 < G < R + 15$.

We performed feature selection using the training-validation set by removing redundant features where the Pearson correlation with more than three other features was above 0.8. After discarding the highly correlated features, we selected (again using only the training-validation set) the 10 features with the highest correlation with the labels. The features selected following this process were `mean_asymmetry`, `compactness_x`, `sat_var`, `avg_hue`, `dom_hue`, `avg_green_channel`, `F1`, `F2`, `F11` and `blue_veil_pixels`.

3.3 Models

We use two models in our experiments: LR and the ResNet-50 CNN. The LR model (scikit-learn [24] implementation) relies on the ten selected standardised handcrafted features. We then use grid search on the training-validation set to find the best C, with inverse regularisation strength parameter $C \in \{0.01, 0.05, 0.1, 0.5, 1, 2, 5\}$.

For the ResNet-50, we use the torchvision `IMAGENET1K_V2` weights, fine-tuned on PAD-UFES-20. We use the ResNet-50 because it was the best performing model in a prior study performed by the dataset authors [19]. We modify the last layer to adapt the model to binary classification. Before training, we resize the images to 224×224 pixels and normalise the intensities according to standard and mean values obtained from ImageNet. We use binary cross-entropy as the loss function and the Adam optimiser with a learning rate of 0.001. We train for 10 epochs with a batch size of 32 (shuffling enabled). Hyperparameter optimisation was not performed due to limited computational resources.

3.4 Experiments

Data Splits. For evaluation, we adopt the setup of [21] with some modifications due to the format of our data. We first split the data into four categories: cancer_female, non_cancer_female, cancer_male and non_cancer_male. We then create five held-out test sets with 104 patients each, consisting of 26 patients from each category. For each test set, all remaining data (including the data from the other test sets) creates a training-validation set. There are no overlaps between each training-validation and test set pair. To avoid data leakage, we kept lesions from the same patient and any augmented lesions in the same set.

We then sample the training-validation sets according to five ratios of female patients: 0%, 25%, 50%, 75%, 100%, creating a total of 25 training-validation sets for each test set, and thus 125 training-validation sets in total[4]. This procedure creates 125 training-validation sets with five held-out test sets for the LR. For the CNN, we use 50 training-validation datasets and two of the five held-out test sets, due to the limitations of our computational resources. (Note that our focus is not on comparing LR vs. CNN performance, but rather to evaluate their respective robustness.) Finally, we evaluate the training ratio impact by comparing the performance on female vs. male patients using the area under the receiver operating characteristic (AUROC) and accuracy (ACC).

Statistical Significance Tests. To test for statistically significant differences across sex ratios, we perform a regression t-test. The null hypothesis is $H_0 : m = 0$, where m is the slope of the linear regression of the ACC or AUROC across all sex ratios. With multiple comparisons, patient sex (female/male), model type (CNN/LR), and evaluation metric (ACC/AUROC), lead to a total of eight (23) tests. We therefore apply a (conservative) Bonferroni correction to avoid Type I errors, leading to a corrected threshold of $\alpha = 0.006$. We then use a Mann-Whitney U-test to test for performance differences between the sexes. Our significance tests are used for assessing the robustness of each model when trained on different female/male patient ratios and then tested on female/male patients. We do not aim to compare LR vs. CNN performance.

4 Results

We show the results for the LR and CNN models in Table 2. Although we cannot directly compare the models, we do not observe large differences in the results,

---

[4] Refer to our Github repository for a figure illustrating the data splits.

with the mean performance for LR being slightly higher than that of the CNN in half of the reported metrics. Our AUROC and ACC scores are lower than, for example, reported by Pacheco et al. [19], however, they also included other clinical factors in their models.

Table 2. Mean and standard deviation of ACC and AUROC for LR and CNN

Metrics	LR	CNN
ACC, f	0.682 ± 0.059	0.650 ± 0.057
ACC, m	0.684 ± 0.080	0.687 ± 0.053
AUROC, f	0.745 ± 0.052	0.716 ± 0.075
AUROC, m	0.727 ± 0.081	0.758 ± 0.059

Figure 1 shows the performance metrics of the LR and CNN, trained on the different sex ratios and then tested separately on female and male patients. We show the results of the statistical significance tests in Table 3. None of the p-values for the regression t-test fall below the corrected threshold, and therefore we consider none of them statistically significant.

We further analyse the performance variability in Fig. 1. For LR, for female patients the lowest standard deviation of the AUROC is at ratio 0.0, steadily

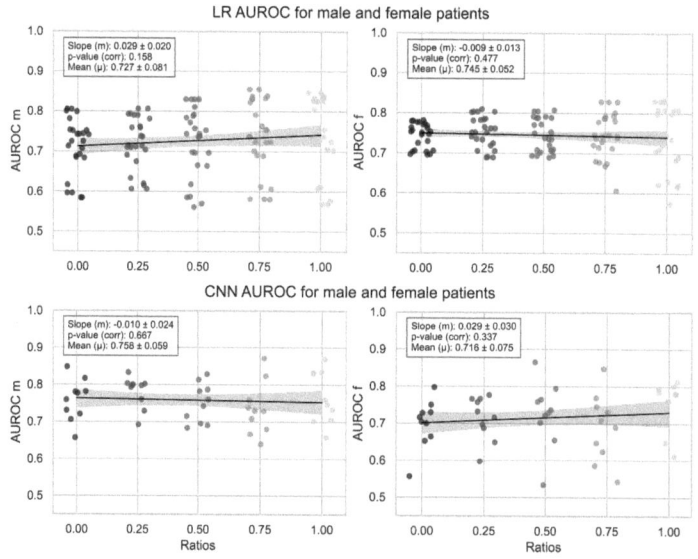

Fig. 1. AUROC evaluated on male (left) and female (right) patients, for LR (top) and CNN (bottom) across different training sex ratios, where a ratio of 1.00 corresponds to a training set entirely of female patients.

Table 3. The *p*-values of the significance tests for LR and the CNN

Statistical test	Threshold	*p*-value
Regression t-test	$\frac{0.05}{8} = 0.006$	
LR AUROC, f		0.477
LR AUROC, m		0.158
CNN AUROC, f		0.3374
CNN AUROC, m		0.6674
Mann-Whitney U test	$\frac{0.05}{4} = 0.0125$	
LR AUROC (f & m)		0.219
CNN AUROC (f & m)		0.007

increasing until ratio 1.0. For male patients we do not observe the same pattern, as the scatter seems consistent across all ratios. In contrast, for the CNN, we see for all plots that the smallest standard deviations are at ratio 0.5.

In Table 3, the Mann-Whitney U-test results show that for the CNN model, there are statistically significant differences in the AUROC between male and female patients, with male patients having higher performance.

5 Discussion and Conclusions

Summary of Results. The results show that for both LR and CNN, the null hypothesis of the models being robust to a data shift change could not be rejected. However, for the CNN, we found that the mean of the ACC and AUROC scores are statistically significantly higher for male than for female patients, which was not the case for the LR: LR is more robust.

Even though we could not reject the null hypothesis, it is worth mentioning that the slope for female patients for the CNN's ACC scores is positive, even when accounting for uncertainty. The slope might have been different if more computational resources had been dedicated to the CNN training. However, in contrast to a previous study [25], we found that for the CNN, the performance was better for male patients in general. We hypothesise that one of the reasons why we found this difference and [25] did not could be because they only processed dermoscopic images, which might have led to less difference between the sexes since the images are generally of a higher quality and the lesions are better visible than in the smartphone images of which PAD-UFES-20 consists.

Limitations. Our results should be interpreted carefully due to the data and models we used. The PAD-UFES images are taken with smartphones, rather than with a dermatoscope used in other datasets, so our results might not generalise to studies on other datasets. We found multiple errors in the PAD-UFES-20 dataset which we then corrected, but we cannot guarantee that all errors are identified. Therefore, our conclusions might not necessarily hold if, for example, the dataset would be audited or relabeled by experts.

Furthermore, although we split the dataset carefully to use as much data as possible, the amount of training data was relatively small. However, we believe that the rich meta-data in PAD-UFES-20 justifies using smaller sample sizes to investigate robustness, as datasets collected in similar conditions would not necessarily be large.

The handcrafted features used by the LR rely on segmentation masks, which were created manually by students without expertise in dermatology. A different segmentation method would result in different feature values. The CNN used the raw images, but it would be interesting to investigate how incorporating the masks during training would affect the performance.

The LR and CNN models did not reach state-of-the-art performance, although this was also not our goal as we focused on robustness and fairness. In future work, it would be worth investigating other popular architectures. Our code is available and reproducible and allows for such extended experiments.

Ethical Considerations. Biases are not just a feature of the distribution in the dataset and often cannot be disentangled from other factors. For example, people with different skin types are likely to have different incidences of skin cancer. Other causes of bias could include medical education being primarily centred around people with lighter skin [3], and a general lack of access to healthcare for patients belonging to racial and ethnic minorities.

We used the term sex in this paper instead of the term gender as the dataset only mentions male and female individuals, and in medical imaging datasets in general, sex is the variable that is often collected. However, in reality, sex is not binary either, and we acknowledge that simply changing the name of the variable might cause problems for individuals whose gender identity does not align with their biological sex.

Finally, deep learning has a significant carbon footprint. While the scale of our project is comparatively small due to the use of traditional methods and a smaller CNN, we believe researchers should be considerate of the carbon footprint of using (larger and larger) models which could still have biases across different patient groups.

Concluding Remarks. We investigated the robustness of a logistic regression model and a ResNet-50 CNN in the context of skin cancer detection in the PAD-UFES-20 dataset. We investigated different ratios of sex compositions in the training data. Our findings show that both classifiers were robust across the different distributions (the CNN less so, but not statistically significantly), which replicates the conclusions of a study on sex differences in Alzheimer's classification [21] and differs from earlier results in chest x-ray diagnosis [17]. However, overall, the CNN had consistently higher performance for male patients. Therefore, even though both models showed robustness to the dataset distribution, the CNN's better performance in male patients suggests a need for further research into the causal factors that influence model biases, both in skin cancer detection specifically and in the broader field of medical image analysis.

For future work in bias assessment and mitigation, our primary recommendation is to consider model choice as a key factor. Simpler models with hand-crafted features are often neglected in current studies in favour of modern deep learning approaches, but both our study and the one by Petersen et al. [21] suggest that simpler, traditional models may perform similarly well and be more robust to dataset shifts and biases. These findings are in line with recent lines of work on model multiplicity [5] and representation learning [15], which suggest that there are often many (superficially) similarly well-performing models with otherwise very distinct characteristics for a given task.

Acknowledgements. We acknowledge the student groups at IT University of Copenhagen for their annotations: Bees (5), Cute Cats Combating Cancer (4), Dragons (5), Elephants (5), Flamingo (5), GroupIT (5), Hedgehawks (5), Iguanas (soaking the sun) (6), Jaguars (5), Koalas (5), Magical Moles (5), Okapi (5), Possum (4) and Queen Snakes (4).

Disclosure of Interests. The authors declare that they have no competing interests.

References

1. Abbasi, N.R., et al.: Early diagnosis of cutaneous melanoma: revisiting the ABCD criteria. JAMA **292**(22), 2771 (2004)
2. Abbasi-Sureshjani, S., Raumanns, R., Michels, B.E.J., Schouten, G., Cheplygina, V.: Risk of training diagnostic algorithms on data with demographic bias. In: Cardoso, J., et al. (eds.) IMIMIC/MIL3ID/LABELS -2020. LNCS, vol. 12446, pp. 183–192. Springer, Cham (2020). https://doi.org/10.1007/978-3-030-61166-8_20
3. Balch, B.: Why are so many black patients dying of skin cancer?. https://www.aamc.org/news/why-are-so-many-black-patients-dying-skin-cancer
4. Banerjee, I., Bhattacharjee, K., Burns, J.L., Trivedi, H., Purkayastha, S., et al.: "Shortcuts" causing bias in radiology artificial intelligence: causes, evaluation and mitigation. J. Am. Coll. Radiol. (2023)
5. Black, E., Raghavan, M., Barocas, S.: Model multiplicity: opportunities, concerns, and solutions. In: 2022 ACM Conference on Fairness, Accountability, and Transparency. FAccT '22. ACM (2022)
6. Brinker, T.J., Hekler, A., Enk, A.H., Berking, C., Haferkamp, S., et al.: Deep neural networks are superior to dermatologists in melanoma image classification. Eur. J. Cancer **119**, 11–17 (2019)
7. Cassidy, B., Kendrick, C., Brodzicki, A., Jaworek-Korjakowska, J., Yap, M.H.: Analysis of the ISIC image datasets: usage, benchmarks and recommendations. Med. Image Anal. **75**, 102305 (2022)
8. Celebi, M.E., et al.: Automatic detection of blue-white veil and related structures in dermoscopy images. Comput. Med. Imaging Graph. **32**(8), 670–677 (2008)
9. Daneshjou, R., Smith, M.P., Sun, M.D., Rotemberg, V., Zou, J.: Lack of transparency and potential bias in artificial intelligence data sets and algorithms: a scoping review. JAMA Dermatol. **157**(11), 1362–1369 (2021)
10. Daneshjou, R., Vodrahalli, K., Novoa, R.A., Jenkins, M., Liang, W., et al.: Disparities in dermatology AI performance on a diverse, curated clinical image set. Sci. Adv. **8**(32) (2022)

11. DermLite: 7-point checklist. https://dermlite.com/pages/7-point-checklist
12. Gichoya, J.W., et al.: AI recognition of patient race in medical imaging: a modelling study. Lancet Digital Health **4**(6), e406–e414 (2022)
13. Groh, M., et al.: Evaluating deep neural networks trained on clinical images in dermatology with the Fitzpatrick 17k dataset. In: 2021 IEEE/CVF Conference on Computer Vision and Pattern Recognition Workshops (CVPRW), pp. 1820–1828. IEEE (2021)
14. Jiménez-Sánchez, A., Avlona, N.R., de Boer, S., Campello, V.M., Feragen, A., et al.: In the picture: medical imaging datasets, artifacts, and their living review. In: Proceedings of the 2025 ACM Conference on Fairness, Accountability, and Transparency, FAccT '25, pp. 511–531. ACM (2025)
15. Kumar, A., Clune, J., Lehman, J., Stanley, K.O.: Questioning representational optimism in deep learning: the fractured entangled representation hypothesis (2025). https://doi.org/10.48550/ARXIV.2505.11581
16. Label Studio Team: Label studio: Open source data labeling tool (2024). https://labelstud.io
17. Larrazabal, A.J., Nieto, N., Peterson, V., Milone, D.H., Ferrante, E.: Gender imbalance in medical imaging datasets produces biased classifiers for computer-aided diagnosis. Proc. Natl. Acad. Sci. **117**(23), 12592–12594 (2020)
18. Madooei, A., Drew, M.S., Hajimirsadeghi, H.: Learning to detect blue–white structures in dermoscopy images with weak supervision. IEEE J. Biomed. Health Inform. **23**(2), 779–786 (2019). https://doi.org/10.1109/JBHI.2018.2835405
19. Pacheco, A.G.C., Krohling, R.A.: The impact of patient clinical information on automated skin cancer detection. Comput. Biol. Med. **116**, 103545 (2020). https://www.sciencedirect.com/science/article/pii/S0010482519304019
20. Pacheco, A.G., Lima, G.R., Salomao, A.S., Krohling, B., Biral, I.P., et al.: PAD-UFES-20: a skin lesion dataset composed of patient data and clinical images collected from smartphones. Data Brief **32**, 106221 (2020)
21. Petersen, E., Feragen, A., da Costa Zemsch, M.L., Henriksen, A., Christensen, O.E.W., Ganz, M.: Feature robustness and sex differences in medical imaging: a case study in MRI-based Alzheimer's disease detection. In: Medical Image Computing and Computer Assisted Intervention (MICCAI), pp. 88–98 (2022)
22. Pham, T.C., Luong, C.M., Hoang, V.D., Doucet, A.: AI outperformed every dermatologist in dermoscopic melanoma diagnosis, using an optimized deep-CNN architecture with custom mini-batch logic and loss function. Sci. Rep. **11**(1), 17485 (2021)
23. Raumanns, R., Schouten, G., Pluim, J.P., Cheplygina, V.: Dataset distribution impacts model fairness: single vs. multi-task learning. In: MICCAI Workshop on Fairness of AI in Medical Imaging, pp. 14–23. Springer, Heidelberg (2024). https://doi.org/10.1007/978-3-031-72787-0_2
24. scikit-learn: scikit-learn: machine learning in python – scikit-learn 1.4.2 documentation. https://scikit-learn.org/stable/
25. Sies, K., Winkler, J.K., Fink, C., Bardehle, F., Toberer, F., et al.: Does sex matter? Analysis of sex-related differences in the diagnostic performance of a market-approved convolutional neural network for skin cancer detection. Eur. J. Cancer **164**, 88–94 (2022)
26. Wen, D., et al.: Characteristics of publicly available skin cancer image datasets: a systematic review. Lancet Digital Health **4**(1), e64–e74 (2022)
27. Winkler, J.K., et al.: Association between surgical skin markings in dermoscopic images and diagnostic performance of a deep learning convolutional neural network for melanoma recognition. JAMA Dermatol. **155**(10), 1135–1141 (2019)

Sex-Based Bias Inherent in the Dice Similarity Coefficient: A Model Independent Analysis for Multiple Anatomical Structures

Hartmut Häntze[1,2,3](✉), Myrthe Buser[2], Alessa Hering[2],
Lisa C. Adams[3], and Keno K. Bressem[3]

[1] Charité - Universitätsmedizin Berlin, 12203 Berlin, Germany
hartmut.haentze@charite.de
[2] Radboudumc, 6525 GA Nijmegen, The Netherlands
[3] Klinikum Rechts der Isar, Technical University of Munich, 81675 Munich, Germany

Abstract. Overlap-based metrics such as the Dice Similarity Coefficient (DSC) penalize segmentation errors more heavily in smaller structures. As organ size differs by sex, this implies that a segmentation error of equal magnitude may result in lower DSCs in female subjects due to their smaller average organ volumes compared to male. While previous work has examined sex-based differences in models or datasets, no study has yet investigated the potential bias introduced by the DSC itself. This study quantifies sex-based differences of the DSC and the normalized DSC in an idealized setting independent of specific models. We applied equally-sized synthetic errors to manual MRI annotations from 50 participants to ensure sex-based comparability. Even minimal errors (e.g., a 1 mm boundary shift) produced systematic DSC differences between sexes. For small structures, average DSC differences were around 0.03; for medium-sized structures around 0.01. Only large structures (i.e., lungs and liver) were mostly unaffected, with sex-based DSC differences close to zero. These findings underline that fairness studies using the DSC as an evaluation metric should not expect identical scores between male and female subjects, as the metric itself introduces bias. A segmentation model may perform equally well across sexes in terms of error magnitude, even if observed DSC values suggest otherwise. Importantly, our work raises awareness of a previously underexplored source of sex-based differences in segmentation performance. One that arises not from model behavior, but from the metric itself. Recognizing this factor is essential for more accurate and fair evaluations in medical image analysis.

Keywords: Fairness · Segmentation · Dice Similarity Coefficient

1 Introduction

The Dice Similarity Coefficient (DSC) is the most commonly used segmentation metric in medical imaging. Together with Intersection over Union it is the default recommendation of frameworks such as Metrics Reloaded [13]. Its pitfalls, such as indifference to multiple classes [18] or size-dependent performance [6,18] are well reported in the literature and, dependent on the task, it can be advisable to pair it with other non-overlap metrics, such as Hausdorff Distance or Normalized Surface Distance [13]. Alternatively, volume-corrected metrics such as the normalized DSC (nDSC) [14,17] have been proposed and demonstrate reduced bias in contexts with substantial target volume variation, such as white matter lesion segmentation [17], although they have yet to see widespread adoption.

The DSC has been widely used to evaluate the fairness of segmentation models. Typically, models are trained on mixed-sex datasets, and DSCs are calculated separately for male and female subgroups. A common assumption is that a fair model should yield comparable DSCs across sexes, which can be controlled by statistical tests. Some studies report higher DSCs for male subjects [9,10], female subjects [1], or equal outcomes for both sexes [3,12,15,16]. When a performance gap is observed, it can be attributed to unbalanced training data or anatomical differences that make one sex more difficult to segment. However, a third possibility is frequently overlooked: the DSC itself may introduce bias related to sex-specific organ size.

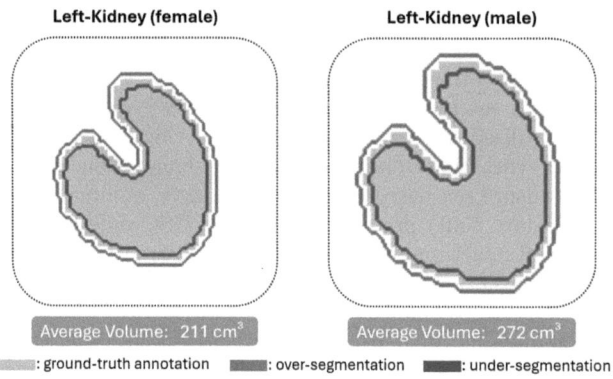

Fig. 1. Example annotation-mask of a left kidney with simulated over- and under-segmentation using a uniform 3 mm margin (grey and black outlines). In our cohort, female kidneys are on average smaller than male ones. Applying the same fixed error systematically across all kidneys in the cohort may result in different DSC values despite an identical error margin, as the increased size of male kidneys positively influences the outcome of the DSC calculation. (Color figure online)

It is well documented that the DSC disproportionately penalizes errors in smaller structures [18], yet few discuss the implications this has on comparisons of subgroups with different organ volumes. Ferrante and Echeveste [7] caution against directly comparing DSC values between pediatric and adult patients for this reason and explicitly raise concerns about sex-based comparisons, given known volume differences between male and female organs [8]. Despite these known limitations of the DSC, their specific impact on sex-based analyses has not been systematically quantified. For example, Häntze et al. [9] reported average DSC scores of 0.89 for male and 0.87 for female subjects, but it remains unclear whether this reflects true model performance differences or biases inherent to the metric itself.

In this study, we explicitly quantify sex-related volume bias in the DSC. Unlike previous work, we do not evaluate the output of a specific segmentation model. Instead, to ensure comparability, we simulate uniform over- and under-segmentation by adding or removing a fixed voxel margin along structure boundaries (Fig. 1). This approach allows us to precisely control the error magnitude and apply identical modifications to subjects of both sexes. Consequently, any observed differences in evaluation metrics cannot be attributed to model behavior or training data composition but must stem from subject-specific characteristics such as sex and organ volume.

2 Methods and Materials

2.1 Study Design

This study is a retrospective simulation. Explicit ethical consent was not required, as the analysis was performed on data from the German National Cohort [2,5] accessed through application number 836.

2.2 Dataset Description

We used a dataset of whole-body in-phase gradient echo sequences of 50 participants (25 male, 25 female) from the German National Cohort with similar age range and BMI (Table 1). The dataset includes annotations for 40 anatomical structures that were quality-controlled by two board-certified radiologists as described in [9]. We excluded left and right femur annotations, as these structures were not fully captured in all sequences. We further removed one annotation of the left kidney that we identified as a pathologic outlier. Additionally, six participants were missing a gallbladder, either due to cholecystectomy or annotation mistakes. A complete list of structures, along with their average volumes, is provided in Table 3. Based on these volumes we categorized the structures into three groups: small (volume $<100\,\text{cm}^3$), medium (100 to $1{,}000\,\text{cm}^3$) and large (volume $>1{,}000\,\text{cm}^3$). We chose these round number thresholds as they are easy to interpret and to apply consistently.

2.3 Evaluation Metrics

We evaluated segmentation quality using the DSC, as it is the most frequently used metric in medical segmentation. Additionally, we used the nDSC due to its proposed advantages; that is robustness against large volumetric variance. The nDSC achieves this by adjusting the DSC by the mean fraction of the positive class for the target structure and group. In this study, this adjustment was

Table 1. Summary statistics (median with interquartile range in parentheses).

Measure	Male (n = 25)	Female (n = 25)
Age	54.0 (14.0)	52.0 (13.0)
Weight (kg)	81.0 (17.0)	64.0 (18.0)
Height (m)	1.78 (0.07)	1.64 (0.09)
BMI	25.69 (5.05)	24.14 (5.42)

applied separately for male and female participants using the mean volume of each anatomical structure by sex. We did not test distance-based metrics such as normalized surface distance or Hausdorff distance. They are, by definition, not affected by volume, hence, errors that we induce with binary dilation will yield the same metrics.

2.4 Experiments

To assess how much the DSC and nDSC differentiate between male and female subjects, we introduced controlled segmentation errors into our ground-truth masks. These synthetic errors are designed to resemble common failures of automated segmentation models, namely over-segmentation and under-segmentation. Because the DSC depends only on the overlap of two shapes, not on the spatial distribution of errors, we can simulate both under- and over-segmentation by applying binary dilation or erosion with a uniform margin around each reference structure, using the scikit-image [19] and Monai [4] python libraries. We resampled the sequences to a isotopic spacing of 1 mm and conducted two experiments: First, to simulate a very good model with almost perfect segmentations we added an error margin for dilation and erosion of 1 mm. Second, to simulate a good but not perfect model, we added an error margin of 3 mm. Note, that due to the isotopic spacing 1 mm equals exactly the width of one voxel. During simulation, we applied both dilation and erosion independently to the reference masks of each participant and structure. We then computed the evaluation metrics for both error types and averaged the results to obtain a single score per structure and participant. To prevent adjacent structures from merging during dilation, all organs were processed independently during both the error simulation and evaluation steps.

2.5 Statistical Analysis

Since organ volumes differ between male and female anatomy, and DSC values are inherently linked to organ size, we expect to find corresponding DSC differences, as this relationship can be derived analytically. Consequently, this article's aim is not to only test whether differences exist, but rather to quantify how large these differences are. For this, we focused on the magnitude of the effects, reporting mean DSC and nDSC differences between male and female participants along with 95% confidence intervals. We performed this analysis for each anatomical structure, across the three volume-based groups, and for both error margins (1 and 3 mm). Within each volume group, we assessed statistical significance using Welch's t-test and controlled the false discovery rate using the Benjamini-Hochberg correction.

3 Results

Applying the 1 mm error margin to the annotations resulted in average DSC values from 0.97 ± 0.00 (both lungs, and liver), up to 0.72 ± 0.10 (right adrenal gland). Table 3 shows the specific DSC values for each structure for this error margin. The 3 mm error margin further reduced the DSC values, which range from 0.92 ± 0.01 up to 0.35 ± 0.10, respectively. The volume corrected nDSC values were generally larger than their DSC counterparts; ranging from 0.98 ± 0.00 (lungs and liver) to 0.83 ± 0.07 (right adrenal gland) for 1 mm and from 0.95 ± 0.00 (right lung) to 0.58 ± 0.04 (left and right iliac arteries) for 3 mm.

The sex-stratified differences for the three volume categories are listed in Table 2. For the 1 mm error margin both sexes had average DSC differences of 0.00 for large, 0.01 for medium and 0.03 for small structures. The nDSC differences are 0.00, 0.00 and 0.04 respectively. The 3 mm margin further increased the differences for both metrics; all differences are significant with adjusted $p < 0.001$. The DSC value distributions by sex for the 1 mm margin are visualized in Fig. 2.

Table 2. Expected metric differences plus 95% confidence intervals for segmentations with an equal error margin between male and female subjects. All differences are significant with adjusted $p < 0.001$.

Category	Error Margin	DSC	nDSC
Large	1 mm	0.00 (0.00, 0.00)	0.00 (0.00, 0.00)
	3 mm	0.01 (0.00, 0.01)	0.00 (0.00, 0.01)
Medium	1 mm	0.01 (0.00, 0.01)	0.01 (0.00, 0.01)
	3 mm	0.02 (0.01, 0.03)	0.02 (0.01, 0.02)
Small	1 mm	0.03 (0.02, 0.04)	0.02 (0.01, 0.02)
	3 mm	0.06 (0.04, 0.08)	0.04 (0.03, 0.05)

4 Discussion

While multiple papers [18] and frameworks, such as metrics reloaded [13], mention the Dice Similarity Coefficient's (DSC) dependence on volume, none discuss the direct implication this has on sex: A subgroup with a larger average organ volume can be expected to have a higher DSC due to inherent bias in the metric itself, independent of model performance.

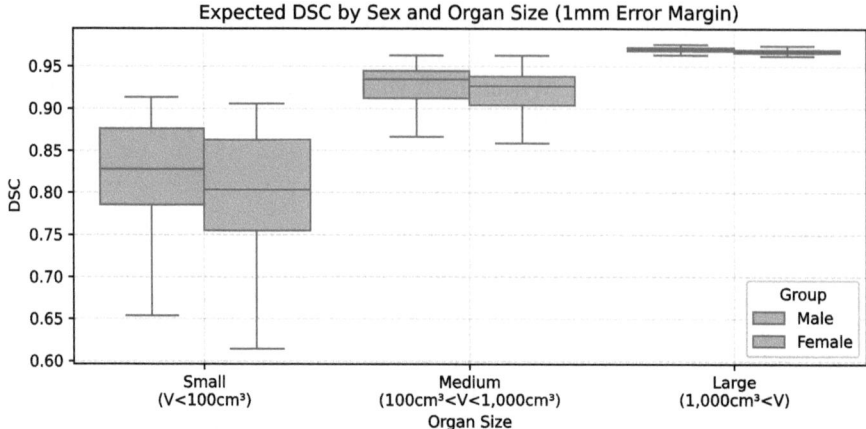

Fig. 2. Box-plot that shows the Dice similarity Coefficient (DSC) for an error margin of 1mm grouped by sex and organ size. Smaller structures are proportionally more affected by volume differences, which results in a higher average DSC difference between sexes. Average differences are 0.03 for small structures, 0.01 for medium structures and 0.00 for large structures.

To quantify this bias we added synthetic segmentation errors to groundtruth annotations and calculated the resulting DSC. Our results demonstrate that even models with high overall accuracy and a small error margin of just 1 mm can produce systematically different DSC values between male and female subjects. For small anatomical structures, we observed an average DSC difference of 0.03; for medium-sized structures, the difference was around 0.01. Only large structures (i.e., lungs and liver) were mostly unaffected by this bias, with DSC differences close to zero (while this difference was statistically significant, we consider it to be of limited practical relevance). When simulating a larger error margin of 3 mm, these discrepancies became more pronounced, with differences reaching up to 0.06 for small structures. In contrast, the normalized DSC (nDSC) was less sensitive towards sex-specific volume differences: for the 1 mm error margin nDSC differences were close to zero for both large and medium structures and reduced to 0.02 for small structures, compared to the standard DSC.

The choice of an appropriate error margin is closely tied to image resolution. For example, with a slice thickness of 3 mm, a 1 mm segmentation error is not feasible due to voxel size constraints. In such cases, the 3 mm error simulations more accurately reflect the impact of a one-voxel segmentation error and are therefore more applicable. Notably, the average DSC values from our 1 mm error simulations closely align with those reported for real-world models in the literature. For instance, Kart et al. [11] reported DSC values of 0.97 (liver), 0.95 (spleen), 0.95 (left kidney), 0.95 (right kidney), and 0.87 (pancreas) on the same cohort. Our simulations yielded comparable values: 0.97, 0.94, 0.94, 0.94, and 0.89, respectively.

Table 3. Average volumes and DSC values for all structures Modifying the ground truth annotations with a uniform 1 mm error margin results in different Dice Similarity Coefficients (DSC) across structures. Column three reports the average DSC values (± standard deviation) across all participants. Column four shows the average DSC differences between male and female subjects, along with the 95% confidence intervals. Volume categories (large, medium, small) are separated by dashed lines.

Class	Avg. Volume	DSC (1mm)	ΔDSC (1mm)
right lung	1948 cm³	0.97 ± 0.00	0.00 [0.00, 0.00]
left lung	1670 cm³	0.97 ± 0.00	0.00 [0.00, 0.00]
liver	1575 cm³	0.97 ± 0.00	0.00 [0.00, 0.00]
small bowel	723 cm³	0.92 ± 0.01	0.01 [0.00, 0.01]
colon	650 cm³	0.92 ± 0.01	0.00 [-0.01, 0.01]
heart	622 cm³	0.96 ± 0.00	0.00 [0.00, 0.01]
right gluteus maximus	568 cm³	0.95 ± 0.01	0.01 [0.01, 0.01]
left gluteus maximus	527 cm³	0.95 ± 0.01	0.01 [0.00, 0.01]
left autochthonous muscle	418 cm³	0.94 ± 0.01	0.01 [0.01, 0.01]
right autochthonous muscle	410 cm³	0.94 ± 0.01	0.01 [0.00, 0.01]
stomach	350 cm³	0.94 ± 0.01	0.01 [0.00, 0.01]
spleen	321 cm³	0.94 ± 0.01	0.00 [0.00, 0.01]
right kidney	275 cm³	0.94 ± 0.01	0.00 [0.00, 0.01]
right hip	262 cm³	0.90 ± 0.01	0.01 [0.01, 0.01]
left hip	261 cm³	0.90 ± 0.01	0.01 [0.00, 0.01]
left iliopsoas muscle	257 cm³	0.92 ± 0.01	0.02 [0.01, 0.02]
left kidney	241 cm³	0.94 ± 0.01	0.01 [0.00, 0.01]
right iliopsoas muscle	241 cm³	0.92 ± 0.01	0.02 [0.01, 0.02]
right gluteus medius	227 cm³	0.94 ± 0.01	0.01 [0.01, 0.01]
spine	221 cm³	0.89 ± 0.01	0.01 [0.01, 0.01]
aorta	221 cm³	0.91 ± 0.01	0.01 [0.01, 0.01]
left gluteus medius	211 cm³	0.94 ± 0.01	0.01 [0.00, 0.01]
urinary bladder	191 cm³	0.94 ± 0.01	0.00 [-0.01, 0.01]
sacrum	170 cm³	0.91 ± 0.01	0.01 [0.00, 0.01]
pancreas	130 cm³	0.89 ± 0.01	0.01 [0.01, 0.02]
gallbladder	123 cm³	0.85 ± 0.05	0.01 [-0.02, 0.04]
inferior vena cava	91 cm³	0.88 ± 0.01	0.01 [0.01, 0.02]
right adrenal gland	86 cm³	0.72 ± 0.10	0.04 [-0.02, 0.10]
left adrenal gland	84 cm³	0.75 ± 0.06	0.05 [0.02, 0.08]
left iliac vena	63 cm³	0.83 ± 0.02	0.02 [0.02, 0.03]
right iliac vena	63 cm³	0.82 ± 0.02	0.03 [0.02, 0.04]
duodenum	63 cm³	0.86 ± 0.03	0.03 [0.01, 0.04]
right gluteus minimus	57 cm³	0.89 ± 0.01	0.01 [0.00, 0.01]
esophagus	55 cm³	0.81 ± 0.02	0.02 [0.01, 0.03]
left gluteus minimus	54 cm³	0.89 ± 0.01	0.01 [0.00, 0.01]
left iliac artery	54 cm³	0.76 ± 0.03	0.04 [0.03, 0.06]
portal vein and splenic vein	50 cm³	0.78 ± 0.02	0.01 [0.00, 0.03]
right iliac artery	44 cm³	0.75 ± 0.04	0.05 [0.03, 0.06]

It is important to understand that our results do not indicate a flaw in the DSC's calculation itself. The metric behaves as intended: the relative impact of a constant-sized error increases as the size of the target structure decreases; and the DSC correctly captures this. However, interpreting such differences as evidence of model unfairness can be misleading, as the absolute error may be identical for both groups. Even a segmentation model that operates in a sex-neutral manner can yield different DSC values between male and female subjects due to inherent anatomical volume differences.

The findings of this study are expected to generalize to other 3D imaging modalities such as CT, as our analysis was based solely on annotations and did not incorporate MRI-specific features. For 2D modalities like X-ray, results are likely to differ as the dimensionality shift from three to two alters the voxel count per structure, and therefore also the calculation of the DSC.

This paper has limitations. Real-world segmentation errors are highly heterogeneous and can include not only over- and under-segmentation but also false positives and complete omissions of structures. By simulating errors through binary dilation and erosion, we simplify this complexity and inevitably lose some similarity to real model behavior. However, the purpose of this study is not to reproduce all error types but to quantify potential DSC differences between male and female participants under idealized conditions. Further, it should be noted that our analysis was conducted on a cohort of German participants, and results may not generalize to populations with different anatomical or sex-based characteristics. Hence, the outcomes presented here should be interpreted as a rough estimate of the potential magnitude of sex-related differences of the DSC. To support comparability, we reported the average organ volumes for our cohort alongside the corresponding metrics (Table 3).

5 Conclusion

In this study, we quantified the expected DSC differences between male and female participants under the assumption of an ideal non-discriminatory segmentation model. Observed DSC differences vary in magnitude from 0.00 to 0.06. This shows that differences in DSC values between sexes do not necessarily indicate model bias. Rather, these discrepancies may be due to a volume-dependent and thus sex-dependent bias intrinsic to the DSC itself, particularly when evaluating small structures below $100\,\mathrm{cm}^3$. Importantly, our results do not aim to replace the DSC, as multiple studies have clearly demonstrated its value in highlighting both negligible [3,12,15,16] and substantial [1,10] sex-based differences. However, we show that systematic differences can arise even under optimal conditions. Consequently, when a segmentation model appears to underperform for one sex, especially for female subjects, it is worth considering whether the metric itself contributes to the observed disparity. Simulations like those presented in this article can help isolate metric-induced effects, and comparing results with alternative metrics such as distance-based metrics or the nDSC [17] may offer a more complete assessment of fairness.

Acknowledgement. This project was conducted with data from the German National Cohort (NAKO) (www.nako.de). The NAKO is funded by the Federal Ministry of Education and Research (BMBF) [project funding reference numbers: 01ER1301A/B/C, 01ER1511D and 01ER1801A/B/C/D], federal states of Germany and the Helmholtz Association, the participating universities and the institutes of the Leibniz Association. We thank all participants who took part in the NAKO study and the staff of this research initiative. Much of the computation resources required for this research was performed on computational hardware generously provided by the Charité HPC cluster. Funded by the European Union. Views and opinions expressed are however those of the author(s) only and do not necessarily reflect those of the European Union or European Health and Digital Executive Agency (HADEA). Neither the European Union nor the granting authority can be held responsible for them.

Disclosure of Interests. The authors have no competing interests to declare that are relevant to the content of this article.

References

1. Afzal, M.M., Khan, M.O., Mirza, S.: Towards equitable kidney tumor segmentation: bias evaluation and mitigation. In: Machine Learning for Health (ML4H), pp. 13–26. PMLR (2023). https://proceedings.mlr.press/v225/afzal23a.html
2. Bamberg, F., et al.: Whole-body MR imaging in the german national cohort: rationale, design, and technical background. Radiology **277**(1), 206–220 (2015). https://doi.org/10.1148/radiol.2015142272
3. de Boer, S., et al.: Robust kidney abnormality segmentation: a validation study of an AI-based framework. arXiv preprint arXiv:2505.07573 (2025)
4. Cardoso, M.J., et al.: Monai: An open-source framework for deep learning in healthcare. arXiv preprint arXiv:2211.02701 (2022)
5. Consortium G.N.C.G: The german national cohort: aims, study design and organization. Eur. J. Epidemiol. **29**(5), 371–382 (2014). https://doi.org/10.1007/s10654-014-9890-7
6. Dice, L.R.: Measures of the amount of ecologic association between species. Ecology **26**(3), 297–302 (1945). https://doi.org/10.2307/1932409
7. Ferrante, E., Echeveste, R.: Open challenges on fairness of artificial intelligence in medical imaging applications. In: Trustworthy AI in Medical Imaging, pp. 265–276. Elsevier (2025). https://doi.org/10.1016/B978-0-44-323761-4.00023-7
8. Geraghty, E.M., Boone, J.M., McGahan, J.P., Jain, K.: Normal organ volume assessment from abdominal CT. Abdom. Imaging **29**, 482–490 (2004). https://doi.org/10.1007/s00261-003-0139-2
9. Häntze, H., et al.: Segmenting whole-body MRI and CT for multiorgan anatomic structure delineation. Radiol. AI e240777 (2025). https://doi.org/10.1148/ryai.240777
10. Ioannou, S., Chockler, H., Hammers, A., King, A.P., Initiative, A.D.N.: A study of demographic bias in CNN-based brain MR segmentation. In: International Workshop on Machine Learning in clinical neuroimaging, pp. 13–22. Springer (2022). https://doi.org/10.1007/978-3-031-17899-3_2

11. Kart, T., et al.: Deep learning-based automated abdominal organ segmentation in the UK biobank and german national cohort magnetic resonance imaging studies. Invest. Radiol. **56**(6), 401–408 (2021). https://doi.org/10.1097/RLI.0000000000000755
12. Lee, T., Puyol-Antón, E., Ruijsink, B., Shi, M., King, A.P.: A systematic study of race and sex bias in CNN-based cardiac MR segmentation. In: International Workshop on Statistical Atlases and Computational Models of the Heart, pp. 233–244. Springer (2022). https://doi.org/10.1007/978-3-031-23443-9_22
13. Maier-Hein, L., et al.: Metrics reloaded: recommendations for image analysis validation. Nat. Methods **21**(2), 195–212 (2024). https://doi.org/10.1038/s41592-023-02151-z
14. Malinin, A., et al.: Shifts 2.0: extending the dataset of real distributional shifts. arXiv preprint arXiv:2206.15407 (2022)
15. Pettit, R.W., Marlatt, B.B., Corr, S.J., Havelka, J., Rana, A.: nnU-Net deep learning method for segmenting parenchyma and determining liver volume from computed tomography images. Ann. Surg. Open **3**(2), e155 (2022). https://doi.org/10.1097/AS9.0000000000000155
16. Puyol-Antón, E., et al.: Fairness in cardiac magnetic resonance imaging: assessing sex and racial bias in deep learning-based segmentation. Front. Cardiovasc. Med. **9**, 859310 (2022). https://doi.org/10.3389/fcvm.2022.859310
17. Raina, V., et al.: Tackling bias in the dice similarity coefficient: introducing NDSC for white matter lesion segmentation. In: 2023 IEEE 20th International Symposium on Biomedical Imaging (ISBI), pp. 1–5. IEEE (2023). https://doi.org/10.1109/ISBI53787.2023.10230755
18. Reinke, A., et al.: Common limitations of image processing metrics: a picture story. arXiv preprint arXiv:2104.05642 (2021)
19. Van der Walt, S., et al.: scikit-image: image processing in python. PeerJ **2**, e453 (2014). https://doi.org/10.7717/peerj.453

The Impact of Skin Tone Label Granularity on the Performance and Fairness of AI Based Dermatology Image Classification Models

Partha Shah[1]([✉]), Durva Sankhe[1], Maariyah Rashid[1], Zakaa Khaled[1], Esther Puyol-Antón[1], Tiarna Lee[1], Maram Alqarni[1], Sweta Rai[2], and Andrew P. King[1]

[1] School of Biomedical Engineering and Imaging Sciences, King's College London, London, UK
partha.shah@kcl.ac.uk
[2] Dermatology Department, Kings College Hospital NHS Foundation Trust, London, UK

Abstract. Artificial intelligence (AI) models to automatically classify skin lesions from dermatology images have shown promising performance but also susceptibility to bias by skin tone. The most common way of representing skin tone information is the Fitzpatrick Skin Tone (FST) scale. The FST scale has been criticised for having greater granularity in its skin tone categories for lighter-skinned subjects. This paper conducts an investigation of the impact (on performance and bias) on AI classification models of granularity in the FST scale. By training multiple AI models to classify benign vs. malignant lesions using FST-specific data of differing granularity, we show that: (i) when training models using FST-specific data based on three groups (FST 1/2, 3/4 and 5/6), performance is generally better for models trained on FST-specific data compared to a general model trained on FST-balanced data; (ii) reducing the granularity of FST scale information (from 1/2 and 3/4 to 1/2/3/4) can have a detrimental effect on performance. Our results highlight the importance of the granularity of FST groups when training lesion classification models. Given the question marks over possible human biases in the choice of categories in the FST scale, this paper provides evidence for a move away from the FST scale in fair AI research and a transition to an alternative scale that better represents the diversity of human skin tones.

Keywords: Bias · AI · Fairness · Dermatology · Granularity

1 Introduction

Artificial intelligence (AI) models based upon deep learning have shown good performance in automatically classifying skin lesions from dermatology images [6]. However, subsequent work has raised concerns about possibly biased or unfair behaviour of these models. For example, an AI model for classifying skin lesions that is trained using data from mostly lighter-skinned patients may perform better on other lighter-skinned patients compared to darker-skinned patients, i.e. its performance will be *biased* in favour of lighter-skinned patients [5,8]. This is concerning because most available databases of dermatology images are either imbalanced by skin tone or do not report skin tone information [4]. Furthermore, when skin tone information is reported, it is common to do so using the Fitzpatrick Skin Tone (FST) scale [7], which classifies skin tone into one of six categories, based on its response to ultraviolet (UV) light. The FST scale is often incorrectly conflated with race [20], is not perfectly correlated with objective assessments of skin tone [21] and has been criticised for its disproportionate focus on lighter skin tones [14,20], i.e. the categories assigned to darker-skinned patients are more "coarse" or less "granular" than those of lighter-skinned patients. Indeed, when the FST scale was first proposed, there were no categories for darker-skinned patients [7]. In other areas of healthcare AI, "too coarse" race information has been shown to have a significant impact on bias assessments [13].

A common approach to mitigate bias in AI models is to train or fine-tune multiple models using (entirely or mostly) protected group-specific data. If protected group information is known at inference time, these protected group-specific models can achieve better performance than a generic model trained with balanced data from all groups. Such improvements have been reported in AI models for cardiac magnetic resonance image segmentation [16], chest X-ray classification [22] and lesion classification from dermatology images [5]. This approach may also be preferable due to the known differences in prevalence and presentation of skin cancer between light- and dark-skinned people [2] - protected group-specific models will have the opportunity to learn group-specific features to optimise their performance. However, when training protected group-specific models it is still necessary to define the protected groups. As outlined above there are question marks over possible human biases in the most popular way of defining protected groups in dermatology, the FST scale.

Therefore, this paper goes beyond the evaluation of protected group-specific models for bias mitigation in dermatology, and evaluates the impact of the granularity of the FST scale used to define these groups. We first investigate the training of baseline protected group-specific models based on three FST scale groups: 1/2, 3/4 and 5/6. We then artificially decrease the granularity of the FST scale groups and assess the effect on model performance for different protected groups. This paper represents the first study into the effect of skin tone label granularity in AI based skin lesion classification.

The paper is organised as follows. Section 2.1 describes the datasets utilised in the experiments. Section 2.2 describes the training of the AI classification models. Section 3 presents the experiments and results, which are then discussed and conclusions drawn in Sect. 4.

2 Materials and Methods

2.1 Datasets and Preprocessing

Two publicly available datasets of clinical dermatology images were utilised: the Diverse Dermatology Images (DDI) [5] and Fitzpatrick 17k [8] datasets.

The DDI dataset contains 656 images with FST scale information and malignant/benign diagnostic labels. The FST scale information is provided as combined categories, i.e. FST types 1 and 2 are combined into a single category, and likewise for types 3/4 and 5/6.

The Fitzpatrick 17k dataset contains 16,577 images along with associated FST scale information (types 1–6) and a 3-class diagnostic label (benign, malignant or non-neoplastic). In this paper, the target task was to classify malignant vs. benign lesions. Therefore, from the full dataset, only images with benign or malignant diagnostic labels were used and images with a non-neoplastic diagnosis were excluded. To make the FST labels consistent with those of the DDI dataset, we labelled each image with a combined FST category, i.e. 1/2, 3/4 or 5/6.

In the experiments described in Sect. 3, the data from the DDI and Fitzpatrick 17k datasets were combined. This was crucial to achieve a sufficiently robust dataset for our analysis, particularly given the reduced sample sizes after excluding data to ensure a clear and clinically meaningful classification task and to obtain adequate representation of darker skin tones. For both datasets, images for which the diagnostic label did not clearly indicate whether the condition was benign or malignant were excluded. For the Fitzpatrick17k dataset, benign cases consisted of conditions such as seborrheic keratosis, dermatofibroma, warts and pilar cysts (mucous cysts were excluded), as well as a number of various nevi such as becker, congenital and halo. Malignant samples consisted of only basal cell carcinoma and squamous cell carcinoma. A similar process was followed for the DDI dataset, where benign samples included blue nevus, dysplastic nevus and melanocytic nevi. This filtering step ensured that the diagnostic categories (i.e. benign/malignant) were used consistently across both datasets. Furthermore, samples with missing FST labels were excluded. After this filtering, we used 1,746 images from the Fitzpatrick 17k dataset (FST 1/2: 1037, FST 3/4: 574, FST 5/6: 135) and 364 images from the DDI dataset (FST 1/2: 136, FST 3/4: 147, FST 5/6: 81).

2.2 Model Architecture and Training

We employed a DenseNet-161 model architecture [10] and trained all models using the Adam optimiser. The maximum number of training epochs was 100. A grid search hyperparameter optimisation was performed separately for each model to optimise batch size (16, 32, 64) and learning rate (0.0005, 0.0003, 0.0001). Model selection (for both hyperparameter choice and for choosing the best epoch) was performed using highest validation accuracy. Following [1], we applied data augmentation based on random rotations (max 90°), width and height shift range of 0.15, vertical and horizontal flipping and brightness changes between 0.8 and 1.1. All images were resized to 224 × 224 pixels before being used for training and evaluation. Models were trained for the binary classification task of malignant vs. benign lesion type.

3 Experiments and Results

We now introduce the experiments we performed to investigate the impact of skin tone label granularity on model performance and bias. To carry out this investigation, we formed a number of subsets of the overall dataset as illustrated in Fig. 1. In each experiment we controlled for training set size, i.e. all compared models were trained with the same number of subjects (165 images). Each training subset was also controlled to ensure approximately equal class balance between benign and malignant lesions. The proportion of benign to malignant lesions was approximately 1:1. When training each model, the training subset was randomly split into 80% for training and 20% for validation. The same test set was used for all experiments, consisting of 50 FST 1/2, 50 FST 3/4 and 50 FST 5/6 images, with a 1:1 benign/malignant ratio.

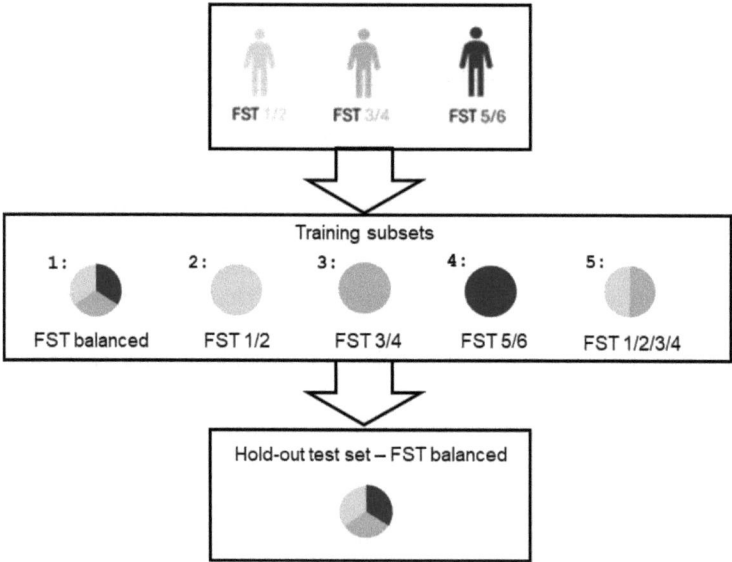

Fig. 1. Illustration of data subsets used for training and testing.

Following [5], we report the area under the receiver operating characteristic curve (AUC) as an evaluation metric for all models. We also report the balanced accuracy (BACC) (i.e. the arithmetic mean of sensitivity and specificity). Finally, we report the Expected Calibration Error (ECE) [9], which is a measure of uncertainty calibration, to provide a broader perspective on model performance.

All experiments were run 20 times with different random seeds and data splits and we report the mean and standard deviation of all metrics over all runs.

3.1 Experiment 1: Baseline Protected Group-Specific Models

The first experiment aimed to determine whether training on FST-specific data, an approach also known as stratified Empirical Risk Minimisation (ERM) [22], improves performance and/or fairness compared to training on mixed (FST-balanced) data (see leftmost four training subsets in Fig. 1). We evaluate all four models using AUC, BACC and ECE on each FST group individually. In addition, we quantify the fairness gap (FG), which represents the difference between the best- and worst-performing FST groups (i.e. a measurement of bias) for each metric. We compute the FG using the mean metric values over the 20 runs for the model trained with FST-balanced data and overall for the approach of training FST group-specific models.

Tables 1, 2 and 3 show the results. We can see that, based on AUC, the use of FST group-specific models improves performance for all three groups (see cells highlighted in italics). The improvement for FST 1/2 was statistically significant using a two-tailed t-test for the AUC, BACC and ECE results. Furthermore, using FST group-specific models reduced the FG for all metrics.

For illustrative purposes, two sample images with their ground truth and model-predicted diagnoses are shown in Fig. 2.

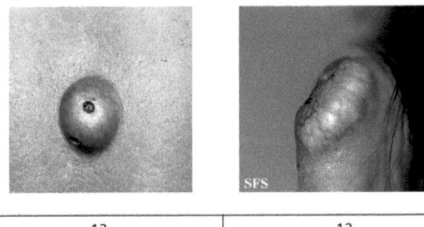

FST:	12	12
Ground truth diagnosis:	Benign	Malignant
Protected group-specific model diagnosis:	Benign	Malignant
Protected group-balanced model diagnosis:	Malignant	Benign

Fig. 2. Example test set images with FST, ground truth diagnosis and diagnoses predicted by FST-specific model and FST-balanced model.

From the results of this first experiment, we conclude that, even when training using data that are balanced by FST group, AI models for skin lesion classification can be biased by skin tone. This bias can be mitigated by training models using FST group-specific data, and as well as fairness improvements, this approach offers some performance improvements too.

Table 1. Experiment 1 - AUC performance of classifier on different data subsets. Results shown stratified by FST group. FST = Fitzpatrick Skin Tone, AUC = area under the receiver operating characteristic curve, FG = fairness gap. Italics indicates higher performance compared to training using FST balanced data (i.e. comparison between the two shaded cells in one column). Bold = p¡0.05 in two-tailed t-test compared to training using FST balanced data.

Training set	Evaluation on test set stratified by FST			
	AUC			
	FST 1/2	FST 3/4	FST 5/6	FG
FST balanced	0.83 ± 0.04	0.86 ± 0.03	0.93 ± 0.03	0.1
FST 1/2	***0.87 ± 0.03***	0.87 ± 0.04	0.88 ± 0.04	
FST 3/4	0.84 ± 0.03	*0.89 ± 0.03*	0.88 ± 0.04	0.06
FST 5/6	0.79 ± 0.03	0.88 ± 0.03	*0.93 ± 0.02*	

Table 2. Experiment 1 - BACC performance of classifier on different data subsets. Results shown stratified by FST group. FST = Fitzpatrick Skin Tone, BACC = balanced accuracy, FG = fairness gap. Italics indicates higher performance compared to training using FST balanced data (i.e. comparison between the two shaded cells in one column). Bold = p¡0.05 in two-tailed t-test compared to training using FST balanced data.

Training set	Evaluation on test set stratified by FST			
	BACC			
	FST 1/2	FST 3/4	FST 5/6	FG
FST balanced	0.76 ± 0.04	0.81 ± 0.04	0.85 ± 0.06	0.09
FST 1/2	***0.79 ± 0.04***	0.80 ± 0.05	0.82 ± 0.05	
FST 3/4	0.78 ± 0.03	*0.83 ± 0.04*	0.81 ± 0.06	0.05
FST 5/6	0.72 ± 0.03	0.83 ± 0.04	0.84 ± 0.06	

Table 3. Experiment 1 - ECE performance of classifier on different data subsets. Results shown stratified by FST group. FST = Fitzpatrick Skin Tone, BACC = balanced accuracy, FG = fairness gap. Italics indicates lower ECE (better calibration) compared to training using FST balanced data (i.e. comparison between the two shaded cells in one column). Bold = p¡0.05 in two-tailed t-test compared to training using FST balanced data.

Training set	Evaluation on test set stratified by FST			
	ECE			
	FST 1/2	FST 3/4	FST 5/6	FG
FST balanced	0.21 ± 0.04	0.16 ± 0.03	0.13 ± 0.05	0.08
FST 1/2	***0.17 ± 0.03***	0.18 ± 0.05	0.17 ± 0.05	
FST 3/4	0.17 ± 0.04	0.16 ± 0.03	0.14 ± 0.03	0.03
FST 5/6	0.25 ± 0.03	0.17 ± 0.04	0.14 ± 0.03	

3.2 Experiment 2: Reducing the FST Granularity of Lighter Skin Tones

The second experiment aimed to assess the impact of reducing the granularity of the FST group labels. We randomly sampled data from the FST 1/2 and FST 3/4 groups to form a single new group (FST 1/2/3/4) with the same training/validation set sizes as the other groups. We trained using the data from this new group and compared the performance of the resulting model on the FST 1/2 and FST 3/4 test data with the performance of the original protected group-specific models (i.e. from Tables 1, 2 and 3). Therefore, in this experiment we compare the second, third and fifth models in Fig. 1.

The results are shown in Tables 4 and 5. Note that the FST 1/2 and FST 3/4 results of these tables are replicated from Tables 1, 2 and 3 and are included to allow easy comparison with the FST 1/2/3/4 results. We can see that the performance of the new 'coarsened' protected group-specific model is always worse than the performance of the original protected group-specific models for all three metrics. For example, when tested on FST 1/2, the original protected group specific model (trained using FST 1/2 data) had a mean AUC of 0.87 and a mean BACC of 0.79; these figures were 0.84 and 0.75 when using the new 'coarsened' model. Both of these differences were statistically significant based on a two-tailed t-test. The performance for the FST 3/4 group also got slightly worse for the 'coarsened' model, but the differences were not significant. Likewise, the ECE values were worse for the 'coarsened' model but the differences were not significant.

From this experiment, we conclude that reducing the FST label granularity when training protected group-specific models can negatively affect performance for some groups.

Table 4. Experiment 2 - AUC and BACC performance of classifier on coarsened subset (i.e. FST 1/2/3/4). Results shown stratified by FST group and compared to results for training using original subsets of FST 1/2 and 3/4 from Tables 1 and 2. FST = Fitzpatrick Skin Tone, AUC = area under the receiver operating characteristic curve, BACC = balanced accuracy. Italics indicates lower performance using coarsened subset compared to original subset (i.e. comparison between the two shaded cells in one column). Bold = p¡0.05 in two-tailed t-test compared to training using FST balanced data.

Training set	Evaluation on test set stratified by FST			
	AUC		BACC	
	FST 1/2	FST 3/4	FST 1/2	FST 3/4
FST 1/2/3/4	**0.84 ± 0.03**	*0.86 ± 0.02*	**0.75 ± 0.04**	*0.79 ± 0.03*
FST 1/2	0.87 ± 0.03	0.87 ± 0.04	0.79 ± 0.04	0.80 ± 0.05
FST 3/4	0.84 ± 0.03	0.89 ± 0.03	0.78 ± 0.03	0.83 ± 0.04

Table 5. Experiment 2 - ECE performance of classifier on coarsened subset (i.e. FST 1/2/3/4). Results shown stratified by FST group and compared to results for training using original subsets of FST 1/2 and 3/4 from Tables 3 and 2. FST = Fitzpatrick Skin Tone, ECE = expected calibration error. Italics indicates higher ECE (worse calibration) using coarsened subset compared to original subset (i.e. comparison between the two shaded cells in one column). Bold = p¡0.05 in two-tailed t-test compared to training using FST balanced data.

Training set	Evaluation on test set stratified by FST	
	ECE	
	FST 1/2	FST 3/4
FST 1/2/3/4	*0.19 ± 0.04*	*0.18 ± 0.03*
FST 1/2	0.17 ± 0.03	0.18 ± 0.05
FST 3/4	0.17 ± 0.04	0.16 ± 0.03

4 Discussion and Conclusions

The FST scale is widely used as a measure of skin tone when assessing bias in AI dermatology models and when attempting to train fairer models [5,8,15]. The main contribution of this work has been to show for the first time that the level of granularity used to record FST scale data can have a significant impact on performance of AI skin lesion classification models for different protected groups.

Specifically, Experiment 1 showed that protected group-specific models (together with inference time knowledge of protected group status) can lead to fairer outcomes and better performance for some groups.

Experiment 2 showed that reducing the granularity of the FST scale when selecting training data generally reduced performance. This is a significant finding, as the FST scale arguably already has higher granularity for lighter skin tones [14], which represents a form of label bias. Therefore, Experiment 2 showed that this type of label bias can negatively impact performance.

We believe that the results we have presented raise question marks over the suitability of the FST scale when training and evaluating fair AI models in dermatology. It should also be noted that the FST scale is not actually a measure of skin tone. Rather, it categorises skin types based on their susceptibility to UV light damage [7]. It follows from this that an individual's FST scale category will never change, but their skin tone may, depending on recent exposure to UV light. This raises further concerns over the suitability of the FST scale in fair AI work. It may be that an alternative scale, such as the Monk skin tone scale [12] or the individual typology angle [3], which do measure skin tone and may be less inherently biased in their choice of categories, will be better suited to use in fair AI research.

Future work could extend the work presented here by evaluating the impact of FST label granularity on algorithmic bias mitigation approaches such as over-/under-sampling [11], loss weighting [11], or Group Distributionally Robust Optimisation (Group DRO) [18]. Additionally, exploring approaches for fair image classification that do not rely on sensitive attributes [17] or focusing on post-processing techniques like those from Ustun et al. for tabular data [19] (which could inspire image-based adaptations), might also be of interest.

Acknowledgements. This research was funded in whole, or in part, by the Wellcome Trust, United Kingdom WT203148/Z/16/Z.

References

1. Bello, A., Ng, S.C., Leung, M.F.: Skin cancer classification using fine-tuned transfer learning of DENSENET-121. Appl. Sci. **14** (2024)
2. Bradford, P.T.: Skin cancer in skin of color. Dermatol. Nurs. **21**, 170–177 (2009)
3. Chardon, A., Cretois, I., Hourseau, C.: Skin colour typology and suntanning pathways. Int. J. Cosmet. Sci. **13**(4), 191–208 (1991)
4. Daneshjou, R., Smith, M.P., Sun, M.D., Rotemberg, V., Zou, J.: Lack of transparency and potential bias in artificial intelligence data sets and algorithms: a scoping review. JAMA Dermatol. **157**(11), 1362–1369 (2021)
5. Daneshjou, R., et al.: Disparities in dermatology AI performance on a diverse, curated clinical image set. Sci. Adv. **8**(32), eabq6147 (2022)
6. Esteva, A., Kuprel, B., Novoa, R.A.J., Ko, S.M.S., Blau, H.M., Thrun, S.: Dermatologist-level classification of skin cancer with deep neural networks. Nature **542**, 115–118 (2017)
7. Fitzpatrick, T.B.: The validity and practicality of sun-reactive skin types I through VI. Arch. Dermatol. **124**(6), 869–871 (1988)
8. Groh, M., et al.: Evaluating deep neural networks trained on clinical images in dermatology with the Fitzpatrick 17k dataset. In: Proceedings of IEEE/CVF Conference on Computer Vision and Pattern Recognition (CVPR), pp. 1820–1828 (2021)
9. Guo, C., Pleiss, G., Sun, Y., Weinberger, K.Q.: On calibration of modern neural networks. In: Proceedings of International Conference on Machine Learning (ICML) (2017)
10. Huang, G., Liu, Z., van der Maaten, L., Weinberger, K.Q.: Densely connected convolutional networks. In: Proceedings of IEEE Conference on Computer Vision and Pattern Recognition (CVPR) (2017)
11. Kamiran, F., Calders, T.: Data preprocessing techniques for classification without discrimination. Knowl. Inf. Syst. **33**, 1–33 (2012)
12. Monk, E.: The Monk skin tone scale (2023). https://doi.org/10.31235/osf.io/pdf4c
13. Movva, R., et al.: Coarse race data conceals disparities in clinical risk score performance. In: Proceedings of Machine Learning for Healthcare (MLHC) (2023)
14. Okoji, U.K., Taylor, S.C., Lipoff, J.B.: Equity in skin typing: why it is time to replace the Fitzpatrick scale. Br. J. Dermatol. **185**(1), 198–199 (2021)
15. Pakzad, A., Abhishek, K., Hamarneh, G.: CIRCLe: color invariant representation learning for unbiased classification of skin lesions. In: Proceedings of European Conference on Computer Vision (ECCV) (2022)
16. Puyol-Antón, E., et al.: Fairness in cardiac MR image analysis: an investigation of bias due to data imbalance in deep learning based segmentation. In: Proceedings of Medical Image Computing and Computer Assisted Interventions (MICCAI), pp. 413–423 (2021)

17. Renggli, C., Smith, D.B., Gola, H.M., Kindermans, P.J.: Do we need training data? Towards fairness in computer vision with no-reference metrics. arXiv preprint arXiv:2309.05148 (2023)
18. Sagawa, S., Koh, P.W., Hashimoto, T.B., Liang, P.: Distributionally robust neural networks for group shifts: on the importance of regularization for worst-case generalization. In: Proceedings of International Conference on Machine Learning (ICML) (2020)
19. Ustun, F.O., Laumann, J.R., Smith, A.D.: Fairness without demographics in repeated decisions. In: Proceedings of the 36th International Conference on Machine Learning (2019). https://proceedings.mlr.press/v97/ustun19a.html
20. Ware, O.R., Dawson, J.E., Shinohara, M.M., Taylor, S.C.: Racial limitations of Fitzpatrick skin type. Cutis **105**(2), 77–80 (2020)
21. Weir, V.R., Dempsey, K., Gichoya, J.W., Rotemberg, V., Wong, A.K.I.: A survey of skin tone assessment in prospective research. npj Dig. Med. **7**(191) (2024)
22. Zhang, H., Dullerud, N., Roth, K., Oakden-Rayner, L., Pfohl, S., Ghassemi, M.: Improving the fairness of chest X-ray classifiers. In: Proceedings of Conference on Health, Inference, and Learning, pp. 204–233 (2022)

Causal Representation Learning with Observational Grouping for CXR Classification

Rajat Rasal[✉], Avinash Kori, and Ben Glocker

Department of Computing, Imperial College London, London, UK
{rrr2417,a.kori21,b.glocker}@imperial.ac.uk

Abstract. Identifiable causal representation learning seeks to uncover the true causal relationships underlying a data generation process. In medical imaging, this presents opportunities to improve the generalisation and robustness of task-specific latent features. This work introduces the concept of grouping observations to learn identifiable representations for disease classification in chest X-rays via an end-to-end framework. Our experiments demonstrate that these causal representations improve performance across multiple classification tasks when grouping is used to enforce invariance with respect to race, sex, and imaging views.

Keywords: Causal representation learning · Invariant representations · Classification · Identifiability

1 Introduction

It is well established that discriminative models trained on chest X-rays (CXRs) from specific demographic groups can inadvertently rely on group-specific patterns, such as trends associated with sex, race, or the imaging modality. These patterns often fail to generalise to other populations, potentially amplifying health disparities [16]. Moreover, it remains unclear whether the presence of group characteristics in learned feature representations are useful for downstream predictive tasks [5].

Identifiable and causal representation learning offers a promising solution to address these challenges. Identifiable models provide theoretical guarantees for learning representations that are consistent across training configurations by recovering the true underlying generative structure of the data [30]. Causal representation learning (CRL), assuming a task-specific data generation process, focuses on learning representations that follow a causal structure which in turn remain invariant across environments (*i.e.*, *populations, groups*) [22,30]. Together, identifiability and causal invariances are critical for building generalisable and trustworthy medical AI systems [3].

Early works in representation learning, particularly disentanglement methods [7], have shown success in controlled synthetic settings with well-defined sources of variation [15] and in some medical applications [14]. However, these approaches typically fall short of achieving identifiability and often struggle to generalise in real-world medical imaging scenarios. Identifiability has been extensively studied in Independent Component Analysis (ICA) [12], where independent features can be recovered up to scaling and permutation. Yet, when observations undergo non-linear mixing, a standard assumption in real-world medical imaging datasets, the recovery of independent latent variables is fundamentally ill-posed [15].

In such real-world datasets, disentangled representations alone often fail to support robust generalisation. Learning the underlying causal structure offers a more promising avenue, as causal representations are invariant across environments, improving both interpretability and predictive reliability [17,22], a necessity in autonomous healthcare systems [18]. CRL specifically aims to uncover the dependency structure within the latent space. Recent advances in CRL leverage invariances and data symmetries to infer identifiable representations from observational data [12,17,30], while other approaches rely on interventional [1] or counterfactual [2] signals to improve identifiability.

In medical imaging, learning identifiable representations offers the potential to improve the generalisation and robustness of task-specific features [3,22]. Recent approaches have explored grouping observations as a strategy to learn invariant causal representations [17]. Building on these ideas, we propose an end-to-end framework for disease classification in CXRs that learn identifiable, causal representations invariant to demographic and non-demographic properties. Our main contributions are as follows:

1. We present an end-to-end image classification framework which learns identifiable representations via causal representation learning (Sect. 2.3).
2. We introduce an observational grouping strategy that enforces invariance to population characteristics, such as race, sex, and imaging view (Algorithm 1).
3. We demonstrate through qualitative and quantitative experiments that our method improves the generalisation and robustness of disease classification across CXR datasets (Sect. 3).

2 Methods

2.1 Background

Our approach to CRL through observational grouping aligns with and enhances existing mutual information-based contrastive learning methods [21,25–27]. These works show that contrastive learning models observational data under non-IID assumptions to extract invariant causal representations. It is important to note, however, that our approach is not contrastive by design; rather, contrastive objectives naturally emerge from the causal assumptions we make [33]. In particular, [25,27] utilises data augmentations for observational grouping to disentangle invariant content features from style features. [21] takes an interesting approach, where the notion of grouping is achieved by generating counterfactual pairs based on desired attributes. In contrast, our work explicitly groups observations based on protected characteristics or other relevant subgroup attributes, thereby inducing non-IID dependencies [17], to jointly learn attribute-invariant representations and tackle a downstream classification task.

2.2 Problem Setup

We consider a dataset that is partitioned into groups based on an observed attribute. Specifically, the set of groups is defined as $G = \{G_k\}_{k=1}^{K}$ where each group G_k has M_k samples: $G_k = \{x_k^i\}_{i=1}^{M_k}$. We assume that our dataset is non-IID, where, in addition to the imposed grouping structure, each group may follow a different underlying distribution. This can arise, for example, when data is collected from different hospitals, imaging devices, or patient populations. We define a feature extractor ϕ that smoothly maps each image x to a feature representation $z = \phi(x) \in \mathbb{R}^N$. Our classifier is defined as $\Phi = (\psi \circ \phi)$, where ψ is a binary classifier mapping z to class labels $y \in \{0, 1\}$ via a linear layer followed by a sigmoid activation. Our goal is for representations z to be invariant to differences across the groups in G; latent features should not be separable based on the group G_k from which the image is drawn. Additionally, representations should support accurate classification. The invariance properties of ϕ follow [30]. Formal definitions are deferred to an extended version of this work.

2.3 Invariance with Observational Grouping for Classification

In the case of CXRs, we define observational groups G based on patient sex, race, or imaging view. Our objective is to learn feature representations that are invariant to these group attributes; in other words, we aim to capture patterns that are shared across all groups while supporting accurate disease classification. We hypothesise that explicitly leveraging group information leads to better, shared representations for a task. In contrast, giving the model more freedom may allow it to learn group-specific modes in its latent space, limiting generalisation [5].

Causal Representations. Following [27] and the setup in Sect. 2.2, a group-invariant feature extractor ϕ leveraged for a discriminative task *implicitly* minimises the following objective:

$$\mathcal{L}_{\text{INV}} = \underbrace{\sum_{k<k'\in[1,K]} \mathbb{E}\left[\|\phi(x_k) - \phi(x_{k'})\|^2\right]}_{\text{Similarity}} - \underbrace{\sum_{k\in[1,K]} H(\phi(x_k))}_{\text{Uniformity}}, \quad (1)$$

where x_k and $x_{k'}$ are sampled from groups G_k and $G_{k'}$. The similarity term encourages representations from different groups to be close, promoting invariance to group-specific variations. The uniformity term maximises entropy of representations, preventing collapse to trivial solutions and ensuring that the learned features remain informative and diverse [31]. Here, the contrastive-style

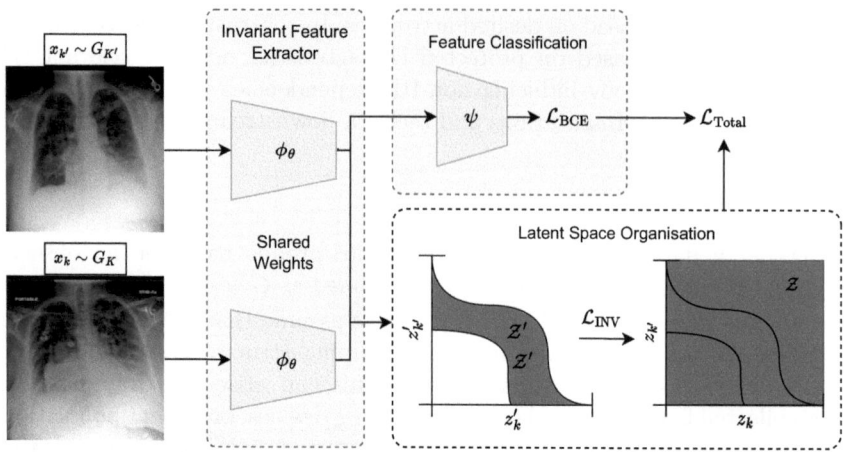

Fig. 1. Training with observational grouping organises the latent space such that representations are invariant to properties across groups in G. We select two groups randomly $k, k' \sim [K]$ and sample images $x_k \sim G_k$ and $x_{k'} \sim G_{k'}$. We use ϕ to extract features, and jointly incorporate them into an invariant loss ($\mathcal{L}_{\text{INV}}$) and a binary classification loss ($\mathcal{L}_{\text{BCE}}$). The invariant loss ($\mathcal{L}_{\text{INV}}$) structures the latent space by relaxing the IID assumption, leading to a transformation from $\mathcal{Z}'$ to $\mathcal{Z}$, as illustrated. Here, the blue coloured regions indicate the theoretical ϕ-supported region of the distribution before and after the use of $\mathcal{L}_{\text{INV}}$. The resulting representation is used for classification with ψ.

objectives naturally emerge from the causal assumptions about the data generation process and grouping structure [33]. Specifically, by treating group membership as soft interventions on latent variables, our method causally identifies invariant content that is shared across groups while discarding group-dependent variations [27].

Training Objective. In this work, we *directly* optimise $\mathcal{L}_{\text{INV}}$ alongside the cross-entropy loss $\mathcal{L}_{\text{BCE}}$ to jointly learn the classifier ψ and invariant feature extractor ϕ:

$$\mathcal{L}_{\text{Total}} = \mathcal{L}_{\text{INV}} + \mathcal{L}_{\text{BCE}}. \tag{2}$$

The end-to-end invariant training procedure is illustrated in Fig. 1, where we sample images x_k and $x_{k'}$ from groups G_k and $G_{k'}$, corresponding to `male` and `female` patients, respectively, both with `pleural effusion`. $\mathcal{L}_{\text{INV}}$ organises the latent space to capture group-invariant features responsible for `pleural effusion`. Intuitively, minimising $\mathcal{L}_{\text{INV}}$ reduces latent intra-class spread across groups, while $\mathcal{L}_{\text{BCE}}$ improves inter-class separation (δ). We hypothesise that this combination not only improves predictive performance but also reduces variance in downstream prediction. We also provide pseudocode for training in Algorithm 1.

Algorithm 1. Data sampling and forward pass for training invariant model.

1: **Input:** Groups $G = \{G_k\}_{k=1}^{K}$; Invariant feature extractor ϕ; Samples per iteration $P \geq 2$; Dataset classes $\mathcal{Y}$. ▷ Groups and labels
2: **Sampling:**
3:      Select class $y \sim \mathcal{U}(\mathcal{Y})$ ▷ Disease selection, $\mathcal{U}$ uniform distribution
4:      $\mathcal{G} = \{G_p \sim \mathcal{U}(G) \mid p \in [1, P]\}$ ▷ Subgroups selection
5:      $\mathcal{X} = \{x_p \sim \mathcal{U}(G_p \mid y) \mid G_p \in \mathcal{G}\}$ ▷ Sampled subgroups filtered by y
6: **Forward Pass:**
7:      $\mathcal{L}_{\text{INV}} = 0, \mathcal{L}_{\text{BCE}} = 0, x_{k'} \sim \mathcal{X}$ ▷ Init. losses and select reference image.
8:      **for** $x_k \in \mathcal{X}$:
9:          $z_k \leftarrow \phi(x_k), z_{k'} \leftarrow \phi(x_{k'})$ ▷ Extract invariant features
10:         $\mathcal{L}_{\text{INV}} \leftarrow \mathcal{L}_{\text{INV}} + \|z_k - z_{k'}\|^2 + H(z_k)$ ▷ Invariance loss
11:         $\mathcal{L}_{\text{BCE}} \leftarrow \mathcal{L}_{\text{BCE}} + \sum \text{BCEWithLogits}(y, \psi(z_k))$ ▷ Classification loss
12:         $x_{k'} \leftarrow x_k$
13:      $\mathcal{L}_{\text{Total}} = \mathcal{L}_{\text{INV}} + \mathcal{L}_{\text{BCE}}$ ▷ Total loss

Identifiability. Following the setup in Sect. 2.2, we can apply the identifiability result from [27]. This guarantees that the invariant content variable $z = \phi(x_k) = \phi(x_{k'})$ is uniquely identified by ϕ. This demonstrates theoretically that the learned representations are stable and reproducible across training configurations, and are robust to distribution shift.

2.4 Data and Implementational Details

Datasets. We use the CHEXPERT [9] and MIMIC [10,11] datasets in this study. We extract a subset containing only images labelled as no findings, pleural effusion or cardiomegaly, with dataset splits included in the extended version of this work. All images are resized to a resolution of 224 × 224. Algorithm 1 details the sampling process used during invariant training. In contrast, the validation and test sampling procedures are the same as the baseline classifiers.

Models and Metrics. Our invariant representation learning strategy is assessed across three deep learning backbones implementing ϕ: ResNet-18 [6], DenseNet-121 [8] and EfficientNetB0 [24]. These models were selected for their strong performance in prior works on CXR analysis [20,23]. By applying our strategy across diverse architectures, we ensure robustness in learning disease-specific features while remaining invariant to confounder such as imaging view (AP, PA), sex (male, female), and race (white, black, asian). We use the area under the receiver operator curve (AUROC) to compare models trained conventionally with $\mathcal{L}_{\text{BCE}}$ against models trained with $\mathcal{L}_{\text{Total}}$ via the invariant strategy.

Training. We train invariant models with two loss components, as seen in Eq. (2); the binary cross-entropy loss $\mathcal{L}_{\text{BCE}}$ and the proposed invariance loss $\mathcal{L}_{\text{INV}}$. We ensure that the gradients from $\mathcal{L}_{\text{INV}}$ update the parameters of ϕ, while $\mathcal{L}_{\text{BCE}}$ updates the parameters of ψ. We apply the same set of hyper-parameters for all models in all our experiments: batch size of 32, learning rate of 0.001 for parameters in ϕ, and learning rate of 0.0001 for parameters in ψ. We learn all models using Adam optimisation for 20 epochs, selecting models with the largest $\mathcal{L}_{\text{BCE}}$ on the validation set for analysis in Sect. 3.

Table 1. AUROC for invariant and non-invariant classifiers.

	CheXpert	MIMIC	CheXpert → **MIMIC**	MIMIC → **CheXpert**
	no findings vs pleural effusion			
ResNet	93.04 ± 0.88	94.31 ± 0.11	90.87 ± 1.51	94.08 ± 0.32
∼ View Inv.	93.78 ± 0.21	93.87 ± 0.32	92.60 ± 0.52	93.78 ± 0.38
∼ Race Inv.	94.23 ± 0.09	94.04 ± 0.08	92.78 ± 0.10	93.97 ± 0.23
∼ Sex Inv.	94.46 ± 0.13	94.30 ± 0.14	93.27 ± 0.29	94.28 ± 0.09
EfficientNet	92.78 ± 0.43	94.62 ± 0.10	91.27 ± 0.85	94.51 ± 0.15
∼ View Inv.	93.46 ± 0.47	94.20 ± 0.12	91.95 ± 1.19	93.88 ± 0.28
∼ Race Inv.	94.32 ± 0.10	94.25 ± 0.10	93.22 ± 0.14	94.15 ± 0.15
∼ Sex Inv.	94.59 ± 0.15	94.46 ± 0.04	93.43 ± 0.24	94.41 ± 0.18
DenseNet	92.78 ± 1.11	93.88 ± 0.73	90.55 ± 1.53	93.81 ± 1.19
∼ View Inv.	94.22 ± 0.11	93.94 ± 0.17	93.24 ± 0.26	93.88 ± 0.18
∼ Race Inv.	94.13 ± 0.40	93.94 ± 0.14	92.69 ± 0.42	93.91 ± 0.20
∼ Sex Inv.	94.31 ± 0.22	94.40 ± 0.12	93.13 ± 0.34	94.32 ± 0.18
	no findings vs cardiomegaly			
ResNet	89.24 ± 1.07	87.27 ± 1.03	87.35 ± 2.40	85.67 ± 2.03
∼ View Inv.	90.56 ± 0.33	91.47 ± 0.49	90.24 ± 0.99	90.02 ± 1.10
∼ Race Inv.	90.76 ± 0.28	91.59 ± 0.25	90.49 ± 0.39	89.48 ± 0.89
∼ Sex Inv.	91.08 ± 0.43	92.40 ± 0.07	90.71 ± 0.55	91.16 ± 0.11
EfficientNet	89.24 ± 1.05	88.23 ± 1.91	87.75 ± 2.23	85.91 ± 2.26
∼ View Inv.	91.07 ± 0.30	91.89 ± 0.10	90.72 ± 0.52	90.28 ± 0.77
∼ Race Inv.	91.27 ± 0.16	92.08 ± 0.10	91.03 ± 0.46	90.19 ± 0.61
∼ Sex Inv.	90.61 ± 0.39	92.41 ± 0.12	89.74 ± 1.24	91.18 ± 0.21
DenseNet	88.41 ± 0.94	87.06 ± 1.42	87.43 ± 1.58	85.21 ± 1.67
∼ View Inv.	90.19 ± 1.42	91.00 ± 0.47	90.75 ± 0.66	90.18 ± 0.40
∼ Race Inv.	90.10 ± 1.54	91.60 ± 0.44	89.64 ± 1.46	89.71 ± 1.38
∼ Sex Inv.	91.24 ± 0.21	92.22 ± 0.12	91.35 ± 0.29	91.08 ± 0.18

3 Experiments

3.1 Identifiability Analysis

We first analyse the latent representations of our model to explicitly evaluate identifiability. For this, we learn every ϕ and dataset combination over 5 random seeds and measure the Mean Correlation Coefficient (MCC) [13] over 1000 latent samples. Across all considered invariances and disease classes, we observe an MCC of around **0.99** with RESNET, **1.0** with EFFICIENTNET, and **0.99** with DENSENET. The high MCC values (near or equal to 1.0) across experiments indicate model convergence in similar local minima irrespective of the starting parameters dictated by random seeds during initialisation. This shows that our representations are stable and unique, crucial for robustness in medical imaging.

3.2 Impact of Invariance Loss

We evaluate the proposed invariance method across three different architectures for ϕ and present results both within and across datasets to assess generalisability. For this analysis, we train invariant and non-invariant models using 10 different random seeds. We select the top 5 models based on validation AUROC and report the mean and standard deviation of their performance. The final tabulated results, summarised in Table 1, provide a comprehensive comparison of the models under different settings. We observe that models trained using the invariance strategy (denoted by the $\sim$ prefix) consistently outperform their respective baseline models, which are listed in the top row of each section. The standard deviation of latent representations in models trained with the invariance strategy is consistently lower than the corresponding baseline, as discussed in Sect. 2.3 and examined further in Sect. 3.3. This supports the hypothesis that invariant training leads to more stable convergence, as suggested by our MCC results in Sect. 3.1. The impact of group sample sizes on model performance is evident, particularly in the case of sex-based invariance, where the larger sample size in each subgroup appears to contribute to higher overall performance.

3.3 Analysis of Invariant Latent Representations

In this section, we analyse the properties of latent representations in models trained conventionally compared to those trained with the proposed invariance strategy. We evaluate this both qualitatively, using principal component analysis (PCA), and quantitatively, using the inter-class separation metric $\delta = \|c_{\text{NF}} - c_{\text{D}}\|_2^2 / s$, where c are first principal component (PC1) of the latent features extracted using ϕ, c_{NF} and c_{D} are the centroids of the **no findings** and disease class clusters, respectively, and $s = \|c_{\text{MAX}} - c_{\text{MIN}}\|_2^2$ scales the distance. Table 2 shows that inter-class separation (δ) is consistently higher for invariant models, and the standard deviation is consistently lower, indicating improved identifiability in line with the MCC metrics in Sect. 3.1. These results provide some intuition for the standard deviations of invariant models being generally lower than those trained conventionally in Table 1. This is visualised in Fig. 2a, which shows the improved linear separability of disease in PC1 with invariant training when compared to PC1 of the conventionally trained model in Fig. 2b.

Table 2. Inter-class separation (δ) for invariant and non-invariant classifiers.

	CheXpert			
	NON INV.	VIEW INV.	RACE INV.	SEX INV.
	no findings vs pleural effusion			
RESNET18	0.245 ± 0.039	0.350 ± 0.011	0.406 ± 0.013	0.404 ± 0.017
EFFICIENTNET	0.379 ± 0.047	0.353 ± 0.037	0.392 ± 0.017	0.414 ± 0.035
DENSENET	0.244 ± 0.017	0.330 ± 0.019	0.339 ± 0.028	0.365 ± 0.025
	no findings vs cardiomegaly			
RESNET18	0.086 ± 0.017	0.271 ± 0.024	0.314 ± 0.013	0.295 ± 0.019
EFFICIENTNET	0.240 ± 0.029	0.293 ± 0.026	0.334 ± 0.011	0.298 ± 0.021
DENSENET	0.034 ± 0.011	0.248 ± 0.019	0.267 ± 0.012	0.277 ± 0.012
	MIMIC			
	no findings vs pleural effusion			
RESNET18	0.281 ± 0.032	0.331 ± 0.012	0.381 ± 0.018	0.377 ± 0.022
EFFICIENTNET	0.334 ± 0.011	0.340 ± 0.027	0.389 ± 0.018	0.41 ± 0.015
DENSENET	0.121 ± 0.063	0.309 ± 0.021	0.342 ± 0.020	0.332 ± 0.016
	no findings vs cardiomegaly			
RESNET18	0.137 ± 0.036	0.297 ± 0.010	0.307 ± 0.018	0.318 ± 0.015
EFFICIENTNET	0.254 ± 0.056	0.294 ± 0.006	0.339 ± 0.019	0.343 ± 0.015
DENSENET	0.020 ± 0.012	0.248 ± 0.015	0.269 ± 0.013	0.276 ± 0.025

4 Discussion and Conclusion

In this work, we investigated grouping-based causal representation learning for disease classification in chest X-rays. By leveraging natural groupings in the data, our method improves the linear separability of disease features, leading to better predictive performance and lower error variability. We support these findings through latent space analysis and by observing consistent convergence behaviours across experiments. Our results, validated on multiple architectures and datasets, show that using our end-to-end invariant training strategy improves model stability and generalisation through identifiability. The choice of grouping attributes was guided by prior research on relevant demographic and non-demographic factors [4,29]. Future work could explore applying this approach to other medical imaging tasks (e.g. scanner invariance in breast imaging), multi-modal data (e.g. text-image pairs), and alternative grouping criteria (e.g. invariance to multiple groups based on assumptions from a causal graph). More broadly, grouping-based methods like ours can be interpreted as enforcing causal invariances to remove spurious shortcuts in classifiers [28]. Enforcing invariances may reduce accuracy or suppress meaningful group differences [19,32], thereby compromising group-fairness. We therefore advise that how and when to use invariance strategies should be guided by domain-specific fairness analyses and

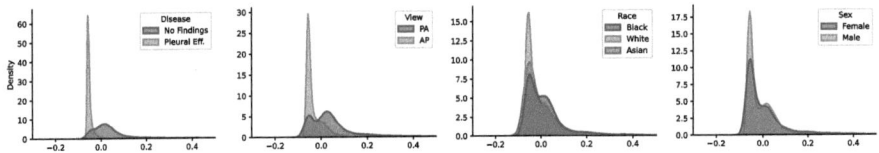

(a) Conventionally trained classifier. Density estimation is performed for disease, view, race and sex (left to right).

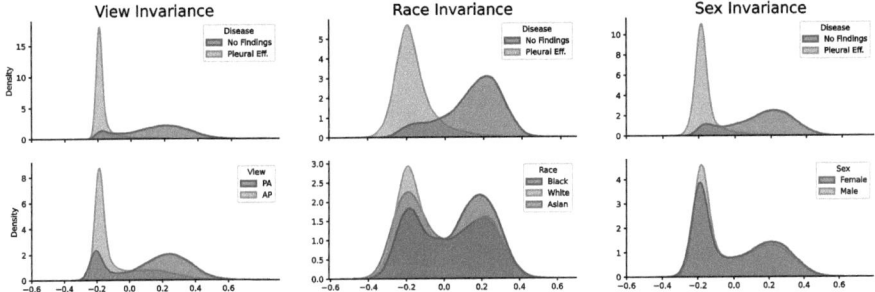

(b) Invariantly trained classifier w.r.t view, race, and sex invariance (left to right). Density estimation is performed for disease (top) and invariant attributes (bottom).

Fig. 2. Density estimation on scaled PC1 embeddings of invariant and non-invariant latent features from ϕ, implemented with a DENSENET backbone, for no findings vs pleural effusion classification.

expert knowledge of the causal data-generating process. Our code is available at https://github.com/RajatRasal/CRL-for-CXR-Classification.

Acknowledgement.. We thank Raghav Mehta, Fabio De Sousa Ribeiro, Pavithra Manoj and Fiona Kekwick for their detailed discussions and insightful feedback on early versions of this manuscript. R.R. is supported by the Engineering and Physical Sciences Research Council (EPSRC) through a Doctoral Training Partnerships PhD Scholarship. A.K. was supported by UKRI (grant no. EP/S023356/1), as part of the UKRI Centre for Doctoral Training in Safe and Trusted AI, and acknowledges support from the EPSRC Doctoral Prize. B.G. received support from the Royal Academy of Engineering as part of his Kheiron/RAEng Research Chair and acknowledges the support of the UKRI AI programme, and the EPSRC, for CHAI - EPSRC Causality in Healthcare AI Hub (grant no. EP/Y028856/1).

Disclosure of Interests. B.G. is a part-time employee of DeepHealth. No other competing interests.

References

1. Ahuja, K., Wang, Y., Mahajan, D., Bengio, Y.: Interventional causal representation learning. arXiv preprint arXiv:2209.11924 (2022)

2. Brehmer, J., De Haan, P., Lippe, P., Cohen, T.S.: Weakly supervised causal representation learning. Adv. Neural. Inf. Process. Syst. **35**, 38319–38331 (2022)
3. Castro, D.C., Walker, I., Glocker, B.: Causality matters in medical imaging. Nat. Commun. **11**(1), 3673 (2020)
4. Gichoya, J.W., et al.: Ai recognition of patient race in medical imaging: a modelling study. Lancet Dig. Health **4**(6), e406–e414 (2022)
5. Glocker, B., Jones, C., Bernhardt, M., Winzeck, S.: Algorithmic encoding of protected characteristics in chest x-ray disease detection models. EBioMedicine **89** (2023)
6. He, K., Zhang, X., Ren, S., Sun, J.: Deep residual learning for image recognition. In: Proceedings of the IEEE Conference on Computer Vision and Pattern Recognition, pp. 770–778 (2016)
7. Higgins, I., et al.: beta-VAE: learning basic visual concepts with a constrained variational framework. In: International Conference on Learning Representations (2017)
8. Huang, G., Liu, Z., Van Der Maaten, L., Weinberger, K.Q.: Densely connected convolutional networks. In: Proceedings of the IEEE Conference on Computer Vision and Pattern Recognition, pp. 4700–4708 (2017)
9. Irvin, J., et al.: Chexpert: a large chest radiograph dataset with uncertainty labels and expert comparison. In: Proceedings of the AAAI Conference on Artificial Intelligence, vol. 33, pp. 590–597 (2019)
10. Johnson, A.E., et al.: Mimic-cxr-jpg, a large publicly available database of labeled chest radiographs. arXiv preprint arXiv:1901.07042 (2019)
11. Johnson, A.E., et al.: Mimic-iii, a freely accessible critical care database. Sci. Data **3**(1), 1–9 (2016)
12. Khemakhem, I., Kingma, D., Monti, R., Hyvarinen, A.: Variational autoencoders and nonlinear ica: a unifying framework. In: International Conference on Artificial Intelligence and Statistics, pp. 2207–2217. PMLR (2020)
13. Khemakhem, I., Monti, R., Kingma, D., Hyvarinen, A.: Ice-beem: identifiable conditional energy-based deep models based on nonlinear ica. Adv. Neural. Inf. Process. Syst. **33**, 12768–12778 (2020)
14. Liu, X., Sanchez, P., Thermos, S., O'Neil, A.Q., Tsaftaris, S.A.: Learning disentangled representations in the imaging domain. Med. Image Anal. **80**, 102516 (2022)
15. Locatello, F., et al.: Challenging common assumptions in the unsupervised learning of disentangled representations. In: International Conference on Machine Learning, pp. 4114–4124. PMLR (2019)
16. Lotter, W.: Acquisition parameters influence AI recognition of race in chest x-rays and mitigating these factors reduces underdiagnosis bias. Nat. Commun. **15**(1), 7465 (2024)
17. Morioka, H., Hyvärinen, A.: Causal representation learning made identifiable by grouping of observational variables. arXiv preprint arXiv:2310.15709 (2023)
18. Pearl, J.: Causality, 2nd edn. Cambridge University Press, Cambridge (2009)
19. Petersen, E., Ferrante, E., Ganz, M., Feragen, A.: Are demographically invariant models and representations in medical imaging fair? arXiv preprint arXiv:2305.01397 (2023)
20. Rajpurkar, P., et al.: Chexnet: radiologist-level pneumonia detection on chest x-rays with deep learning. arXiv preprint arXiv:1711.05225 (2017)
21. Roschewitz, M., de Sousa Ribeiro, F., Xia, T., Khara, G., Glocker, B.: Counterfactual contrastive learning: robust representations via causal image synthesis. In: MICCAI Workshop on Data Engineering in Medical Imaging, pp. 22–32. Springer, Heidelberg (2024). https://doi.org/10.1007/978-3-031-73748-0_3

22. Schölkopf, B., et al.: Toward causal representation learning. Proc. IEEE **109** (2021)
23. Sellergren, A.B., et al.: Simplified transfer learning for chest radiography models using less data. Radiology **305**(2), 454–465 (2022)
24. Tan, M., Le, Q.: Efficientnet: rethinking model scaling for convolutional neural networks. In: ICML, pp. 6105–6114. PMLR (2019)
25. Tosh, C., Krishnamurthy, A., Hsu, D.: Contrastive learning, multi-view redundancy, and linear models. In: Algorithmic Learning Theory, pp. 1179–1206. PMLR (2021)
26. Tsai, Y.H.H., Wu, Y., Salakhutdinov, R., Morency, L.P.: Self-supervised learning from a multi-view perspective. arXiv preprint arXiv:2006.05576 (2020)
27. Kügelgen, J., et al.: Self-supervised learning with data augmentations provably isolates content from style. Adv. Neural. Inf. Process. Syst. **34**, 16451–16467 (2021)
28. Wang, J., Jabbour, S., Makar, M., Sjoding, M., Wiens, J.: Learning concept credible models for mitigating shortcuts. Adv. Neural. Inf. Process. Syst. **35**, 33343–33356 (2022)
29. Yang, Y., Zhang, H., Gichoya, J.W., Katabi, D., Ghassemi, M.: The limits of fair medical imaging AI in real-world generalization. Nat. Med. **30**(10), 2838–2848 (2024)
30. Yao, D., Rancati, D., Cadei, R., Fumero, M., Locatello, F.: Unifying causal representation learning with the invariance principle. arXiv:2409.02772 (2024)
31. Yao, D., et al.: Multi-view causal representation learning with partial observability. arXiv preprint arXiv:2311.04056 (2023)
32. Zhao, H., Dan, C., Aragam, B., Jaakkola, T.S., Gordon, G.J., Ravikumar, P.: Fundamental limits and tradeoffs in invariant representation learning. J. Mach. Learn. Res. **23**(340), 1–49 (2022)
33. Zimmermann, R.S., Sharma, Y., Schneider, S., Bethge, M., Brendel, W.: Contrastive learning inverts the data generating process. In: International Conference on Machine Learning, pp. 12979–12990. PMLR (2021)

Invisible Attributes, Visible Biases: Exploring Demographic Shortcuts in MRI-Based Alzheimer's Disease Classification

Akshit Achara[(✉)], Esther Puyol Anton, Alexander Hammers, Andrew P. King, and for the Alzheimers Disease Neuroimaging Initiative

School of Biomedical Engineering and Imaging Sciences, King's College London, London, UK
akshit.achara@kcl.ac.uk

Abstract. Magnetic resonance imaging (MRI) is the gold standard for brain imaging. Deep learning (DL) algorithms have been proposed to aid in the diagnosis of diseases such as Alzheimer's disease (AD) from MRI scans. However, DL algorithms can suffer from shortcut learning, in which spurious features, not directly related to the output label, are used for prediction. When these features are related to protected attributes, they can lead to performance bias against underrepresented protected groups, such as those defined by race and sex. In this work, we explore the potential for shortcut learning and demographic bias in DL based AD diagnosis from MRI. We first investigate if DL algorithms can identify race or sex from 3D brain MRI scans to establish the presence or otherwise of race and sex based distributional shifts. Next, we investigate whether training set imbalance by race or sex can cause a drop in model performance, indicating shortcut learning and bias. Finally, we conduct a quantitative and qualitative analysis of feature attributions in different brain regions for both the protected attribute and AD classification tasks. Through these experiments, and using multiple datasets and DL models (ResNet and Swin-Transformer), we demonstrate the existence of both race and sex based shortcut learning and bias in DL based AD classification. Our work lays the foundation for fairer DL diagnostic tools in brain MRI. The code is provided at https://github.com/acharaakshit/ShortMR.

Keywords: Bias · Fairness · Shortcuts · Brain · MRI

1 Introduction

Deep learning (DL) algorithms can suffer from shortcut learning, in which spurious correlations between input features and output labels are learnt, regardless of the features' actual relevance to the task [8,21]. When the features are associated with protected attributes such as race and sex these shortcuts can lead to bias. The biases can be further exacerbated by imbalanced representation of protected groups in datasets used for training DL models [1,3]. Datasets such as Waterbirds [26] and CelebA [19] for computer vision, and CheXpert [13] in medical imaging have been created for investigating such spurious correlations.

Spurious features can be learned by DL algorithms for tasks such as classification [23,34], segmentation [22] or generation [7,10]. For datasets such as Waterbirds [26], it is visually possible to identify the spurious correlations between the background (land and sea) and objects (land and sea birds). However, in many medical images, it is difficult or impossible for humans to identify the associated protected attribute(s), such as race and sex, i.e. it could be said that they are "invisible" to humans. However, DL algorithms can often detect them [9,30]. If a DL algorithm can distinguish between different protected groups, then there is the potential for features associated with these attributes to act as shortcuts, or to introduce "visible" biases in clinical classification tasks. Therefore, in this paper, we consider the question: "can protected attribute based shortcut learning lead to bias in DL classification models for brain MRI scans?"

We first study if demographic attributes, namely race and sex, can be identified from structural 3D brain MRI scans on three different datasets. Second, we create baseline and "biased" datasets with differing levels of race/sex imbalance to study shortcut learning in AD classification from brain MRI. Finally, we perform a quantitative and qualitative analysis of regional feature attributions to reveal the nature of the shortcut learning and lay the foundations for future research on understanding the effect of these "invisible" protected attributes on "visible" biases.

Related Work. In [30], the authors showed that sex can be classified from structural adolescent brain MRIs. However, the focus of this work was on assessing sex classification fairness rather than fairness in diagnostic tasks. Previous studies have shown the effect of protected group representation (sex) for AD classification on performance [2,16,24,32]. However, the datasets were not specifically curated for assessing shortcut learning. In [29], the authors studied the effects of various synthetic biases on model performance using a similar strategy of dataset curation. However, the approach utilised synthetic data and synthetically introduced biases with a different research objective.

Contributions. We show that race and sex can be identified from 3D brain MRI scans on three different datasets using two different DL models (ResNet [12] and SwinTransformer [18]). We construct baseline and biased datasets based on sex and race to highlight shortcut learning in DL based AD classification. Finally, we quantitatively and qualitatively analyse feature attributions to show that both sex and race can cause shortcut learning and bias in AD classification, as well as provide insight into the nature of the shortcuts.

2 Methods and Experiments

Notation. We use the following notation throughout this paper:

- $X \in \mathbb{R}^{N \times 1 \times D \times H \times W}$: MRI input images ($N$: number of samples; single channel; D, H, W: depth, height, width). $X_i \in \mathbb{R}^{1 \times D \times H \times W}$ represents a single input image, $i \in \{1, .., N\}$.
- $A \in \{0,1\}^N$: Binary protected attributes (e.g., sex, race) with A_0, A_1 denoting distinct protected groups (e.g., male/female, Black/White).
- $Y \in \{0,1\}^N$: Diagnostic labels with Y_0, Y_1 denoting specific diagnostic classes (i.e. CN: Cognitively normal, AD: Alzheimer's disease).
- f: Predictive classifier; either $f : X \to A$ (protected attribute classification) or $f : X \to Y$ (diagnosis).
- $T : X \to X'$ indicates a non-linear transform in image space.
- $L \in \mathbb{R}^{N \times 1 \times D \times H \times W}$ indicates the positive GradCAM attributions for X.
- $\Omega \in \{1, .., \omega\}^{1 \times D \times H \times W}$ represents an atlas map with ω regions.
- $R : \mathbb{R}^\omega \to \{1, ..., \omega\}$ represents the attribution rank of each region in the atlas.

Datasets

1. **ADNI**: The Alzheimer's Disease Neuroimaging Institute (ADNI) [25] has released several data studies called *ADNI 1, ADNI GO* and *ADNI 2* which consist of structural MRIs, acquisition parameters and protected attributes associated with the subjects. We perform skull stripping using HD-BET [14] on the already-preprocessed images for all our experiments. Each subject may have one or more images. This dataset was used in Experiment 1 on protected attribute classification and Experiment 2 on shortcut learning.
2. **OASIS-3**: The Open Access Series of Imaging Studies 3 (OASIS-3) [17] dataset consists of structural MRIs and protected attribute information for a combination of healthy subjects and subjects at different stages of cognitive decline. We select the bias-normalised images produced by Freesurfer [6] and perform skull stripping on these images. Each subject may have one or more images. This dataset was used only in Experiment 1 on protected attribute classification.
3. **HCP**: The Human Connectome Project (HCP) [31] consists of 1114 structural MRIs from healthy subjects along with protected attribute information. We use the already preprocessed and skull-stripped images available with the dataset. Each subject has one sample in this dataset. This dataset was used only in Experiment 1 on protected attribute classification.

All images were retained in native space and resized to $256 \times 256 \times 256$. Inputs were z-score normalised. Analyses were limited to adult MRIs from Black and White subjects due to limited data from other racial groups.

Experiment 1: Protected Attribute Classification. The first experiment aimed to train and evaluate DL models for race and sex classification. Each classification task is binary, i.e. male/female and Black/White. Formally, the protected attribute classification task can be defined as: $f : X \to A$. We use a stratified train-test ratio of 80 : 20 for both race and sex classification tasks with validation sets being 8% and 12% of the training sets respectively. A smaller validation set is used for race classification due to the class imbalance. Stratification is based on race, sex, and age, with subjects categorized as 'younger' or 'older' using a single age threshold based on absolute age range for each dataset. One model was trained on each dataset.

Experiment 2: Shortcut Learning. In Experiment 2, we consider the CN vs. AD classification task, which can be formally defined as: $f : X \to Y$, where $Y \in \{0, 1\}^N$ are the diagnostic labels (i.e. CN and AD). We created two types of dataset: "baseline" and "biased", with chosen levels of (im)balance between protected attributes (A_0, A_1) and diagnostic labels (Y_0, Y_1). In a baseline dataset, the combinations of $A \in A_0, A_1, Y \in Y_0, Y_1$ are proportionally represented for all the samples (X, Y), preserving their overall distribution within each diagnostic class. In a biased dataset, the training set consists of majority samples (X, Y) such that $A = A_1, Y = Y_0$ or $A = A_0, Y = Y_1$ and minority samples (X, Y) such that $A = A_0, Y = Y_1$ or $A = A_1, Y = Y_0$. Here, A_0 and A_1 represent the two protected groups and Y_0 and Y_1 could be any of the two classes (CN or AD). The test set has an opposite notion of majority and minority samples, i.e. the majority and minority samples in the training set and the minority and majority samples in the test set. This imbalance in samples based on protected groups introduces a group-based distributional shift between training and test datasets.

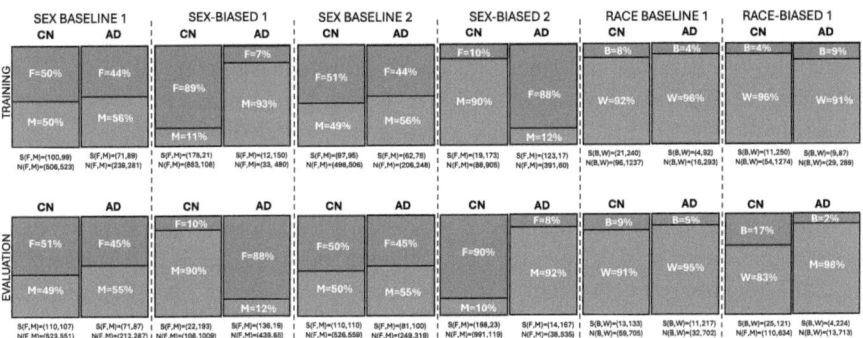

Fig. 1. The formation of baseline and biased datasets. For the baseline datasets, the sex or race (im)balance between training and test sets are approximately the same. In the biased datasets they are different. Here, $S(A_0, A_1)$ indicates the number of subjects for each protected group A_0 and A_1. Similarly, $N(A_0, A_1)$ indicates the number of samples (i.e. images) for each protected group. M, F, B and W correspond to male, female, Black and White

The proportions of samples in the baseline and biased training and test datasets for each experiment can be seen in Fig. 1. The proportions of AD and

CN subjects between the baseline and biased datasets are constrained to be equal (with a minor floating point error) to ensure fair comparison of models. For both balanced and biased datasets, we use a stratified validation set which is comprised of 10% of the training data. The stratification is performed based on cognitive state, race, sex and age ('younger' and 'older'). It is to be noted that there is a less severe difference in protected attribute imbalance between the race baseline and biased datasets due to the insufficient number of Black subjects. Additionally, we have not created a race biased dataset 2 due to lack of sufficient training data.

Experiment 3: Interpretability. We produce feature attribution maps to visualise the regions of the MRI scans that are used by the models in making their classifications. The interpretability method we use is GradCAM [27] which is a layer based attribution method. We use stability certification [15] to analyse the stability of feature attributions. This computes the probability that the prediction remains unchanged when the input with only patches consisting of top features (obtained from any interpretability method) is perturbed by up to r features where r is referred to as the radius. To enable quantification of attributions for different brain regions, the L are transformed to L' using T, which was computed by registering the X_i's to the Hammersmith atlas [11] via image registration. This is followed by obtaining the mean attribution rank vector over all N test samples, $\mathbf{r} = \frac{1}{N}\sum_{i=1}^{N} R(\{\frac{\sum L'_i \cap \Omega_1}{|\Omega_1|}, ..., \frac{\sum L'_i \cap \Omega_\omega}{|\Omega_\omega|}\})$. Here, L'_i represents the GradCAM attribution map for sample i in the atlas space, Ω_j represents all the pixels in region j of the atlas and $L'_i \cap \Omega_j$ represents the pixels in L'_i that fall in Ω_j. Effectively, we rank the mean attributions of the atlas regions for each sample, and then average the rank vectors over all samples.

Feature attribution maps and ranking vectors are produced in this way for both protected attribute classification models ($\mathbf{r}^{PA}$) and the biased and baseline diagnosis models ($\mathbf{r}^{BI}$, $\mathbf{r}^{BA}$). We then compute difference vectors $\mathbf{B} = \mathbf{r}^{BI} - \mathbf{r}^{BA}$ and $\mathbf{P} = \mathbf{r}^{PA} - \mathbf{r}^{BA}$ to find the regions most associated with bias and protected attributes, respectively. To quantify the extent of the shortcut learning we compute the Spearman's rank correlation coefficient between these two vectors. To reveal the regions most associated with shortcut learning we visualise the most significant shared regions netween $\mathbf{B}$ and $\mathbf{P}$ in the atlas space. We use one sample per subject for these interpretability experiments. Overall, this approach is used to understand the relation between features from different models in the atlas space. These rank correlations can be interpretable metrics that provide insights into the occurrence of shortcut learning as well as quantifying its impact.

Implementation Details. We use two DL models in our experiments, namely SwinTransformer (with Convolutional Layers from MONAI [4] and a window size of 5) and ResNet50. Pretrained weights [5] for ResNet50 were utilised for faster convergence due to the relatively small training datasets. Validation performance was used for early stopping mechanisms and obtaining best checkpoints based on validation set loss and F1-scores. It was ensured for all tasks that there was no subject overlap between the training, validation and test subsets. We used inverse probability weighting along with Cross Entropy [28] as our loss function and a cosine learning rate scheduler with the AdamW [20] optimiser. We used a batch size of 4 for all tasks along with gradient accumulation. The code for the implementation is provided at https://github.com/acharaakshit/ShortMR.

3 Results

Experiment 1: Protected Attribute Classification. Table 1a shows that both SwinTransformer and ResNet50 models have high F1-scores in classifying both male and female subjects. Similarly, in Table 1b, even with severe data imbalance, it can be seen that both models can classify White and Black subjects with a high level of accuracy.

Table 1. Experiment 1 - Protected attribute classification. Performance measured by F1 score for SwinTransformer and ResNet50 models, for both race and sex classification across three datasets. The numbers in the cell (a,b) indicate the number of subjects in the complete and test datasets repsectively.

(a) Sex classification

Dataset	Class	SwinTransformer	ResNet50
ADNI	Female (723, 149)	0.87	0.91
	Male (922, 184)	0.89	0.93
OASIS-3	Female (722, 145)	0.93	0.93
	Male (577, 115)	0.91	0.91
HCP	Female (543, 108)	0.91	0.94
	Male (455, 92)	0.88	0.92

(b) Race classification

Dataset	Class	SwinTransformer	ResNet50
ADNI	White (658, 132)	0.98	0.98
	Black (85, 17)	0.81	0.77
OASIS-3	White (1101, 221)	0.98	0.99
	Black (198, 39)	0.84	0.91
HCP	White (830, 167)	0.98	0.97
	Black (168, 33)	0.93	0.84

Experiment 2: Shortcut Learning. Figure 2 shows the differences in class-level F1-scores between the baseline and biased models for the CN vs. AD task. A steeper drop is observed in the AD class as compared to CN. The second row highlights the differences in group-level accuracy scores between the baseline and biased models, depicting the occurrence of shortcut learning.

Experiment 3: Interpretability. For ResNet50, the Spearman's rank correlations ρ between R(**B**) and R(**P**) (see Sect. 2) were 0.85, 0.59 and 0.32 for the Sex 1, Sex 2 and Race 1 experiments, respectively (see Fig. 1). Similarly, ρ was 0.66, 0.67 and 0.08 for the Sex 1, Sex 2 and Race 1 experiments, respectively for SwinTransformer. All p values were < 0.001, indicating significant correlations except for the Race 1 experiment for SwinTransformer ($p = 0.41$), where we did not see a correlation. We also performed a permutation test to account for dependence of **P** and **B** on $\mathbf{r}^{BA}$. We observed significant p-values (< 0.05) for $S1$ using ResNet50 and $S1, R1$ using SwinTransformer.

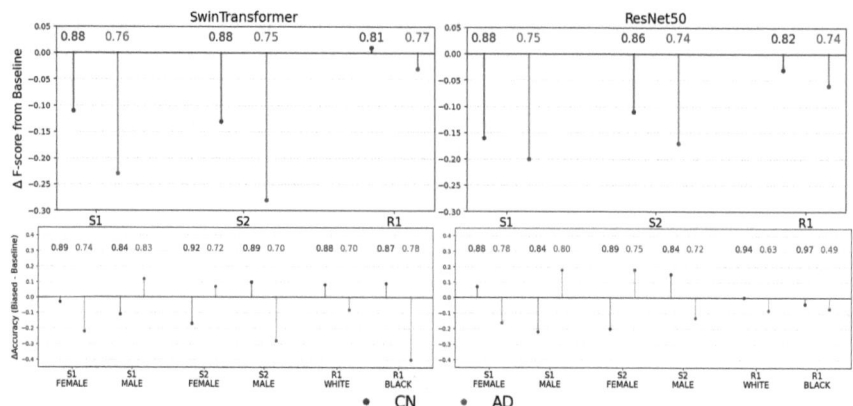

Fig. 2. Experiment 2 - Shortcut learning. Figures show differences in performance (F1-score and accuracy) between models trained on baseline CN vs. AD task (from ADNI) and the same task with the corresponding biased datasets. $S1$, $S2$ and $R1$ represent the sex biased 1, sex biased 2 and race biased 1 datasets respectively (see Fig. 1). The number above each line indicates the baseline F1 score/accuracy, and a positive/negative Δ indicates an increase/drop in F1 score/accuracy from the baseline to the biased model. The first row shows overall class-level F1 scores and the second row shows protected group level accuracies

Next, we establish the stability of our explanations. We selected the top 12.5% patches with highest attributions and added brain patches with an increasing radius. We observed an average soft stability (computed over two different models) of about 0.95 for all radii (with steps) up to 90 (patches) for the ResNet50 model indicating high stability. For SwinTransformer, we observed an overall lower average stability of about 0.79 and relatively more deviations. Therefore, in Fig. 3, we show the qualitative results for ResNet50 only for a more reliable interpretability analysis.

4 Discussion

Our work has shown that protected attributes (race and sex) can be classified to a high level of accuracy across multiple datasets and with two different DL models, one CNN based and one Vision Transformer based. This is an important prerequisite for the presence of bias in DL models and furthermore has the potential to lead to protected attribute based shortcuts. Given that many public brain MRI datasets do not report protected attribute information or are imbalanced, this raises concern for the use of DL for clinical classification purposes.

Our baseline CN vs. AD classification model showed strong performance. However, this performance dropped significantly for the sex biased datasets. The test performance on majority classes (in training) generally increased and generally decreased from the baseline models on the minority classes. There was

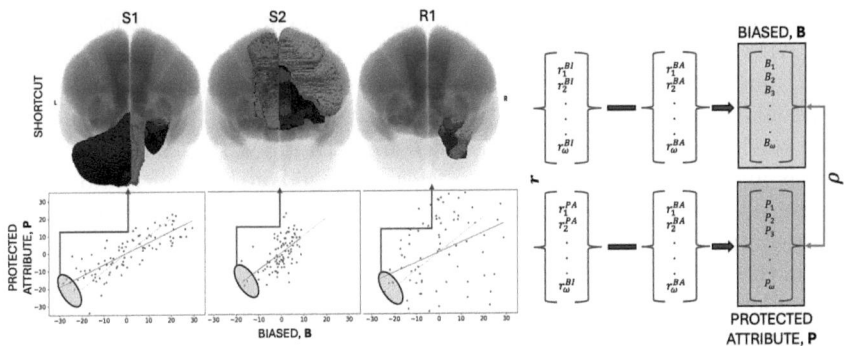

Fig. 3. Experiment 3 - Interpretability. $S1$, $S2$ and $R1$ highlight the top-5 regions (shortcut features) with the highest biased + protected attribute ranks ($\mathbf{B}+\mathbf{P}$). Negative indicates a higher rank and a darker shade of colour indicates more importance. The plots illustrate the linear relationships between biased and protected attribute ranks

also a drop for the race biased model, although it was less pronounced. This is likely due to the less severe difference in imbalance in White and Black subjects between the training and test datasets. This was unavoidable as the total number of Black subjects (CN+AD) was relatively low. Additionally, the SwinTransformer model was trained from scratch, whereas ResNet50 used pre-trained weights, and this could have caused a steeper drop in Black AD classification due to a distributional shift. Essentially, due to the low number of samples from Black subjects with AD, the SwinTransformer model may not have been able to classify them correctly (see Fig. 2). However, a deeper analysis is required with more Black subjects to produce stronger evidence of shortcut learning and/or distributional shifts.

Qualitatively, it was initially difficult to find common patterns across models and tasks that could provide conclusive evidence that the biased models were utilising features related with the protected attributes. This is in contrast to commonly seen feature attributions for models trained on datasets like Waterbirds, where shortcut features like Background are clearly seen. Therefore, we performed rank based analysis to highlight the regions contributing to the shortcut learning. This novel analysis enabled us to quantify the extent of the shortcut learning as well as reveal the regions that were associated with it, even though they are not "visible" to humans. This approach could be used to gain a deeper understanding of shortcut learning in other applications in medical imaging.

In summary, our work has shown that sex and race can be classified from 3D adult brain MRIs with high accuracy, representing the most comprehensive study of protected attribute identification from brain MRI to date. Sex classification (on adolescent brain MRIs) has previously been demonstrated in [30] but to the best of our knowledge, this is the first time that race classification from brain MRI has been investigated. Furthermore, we have shown that protected attribute based (im)balance in the training set can cause shortcut learning and bias in AD classification. This is consistent with the findings of [32], who found that DL models for AD classification suffered from bias when trained with mixed (but imbalanced by sex and race) datasets. Additionally, another study [33] showed that age and sex can be classified from Brain MRIs resulting in biased diagnosis and that augmentation strategies may aid in bias mitigation. All of these works examined bias when training with mixed (but imbalanced) training sets. In contrast, in our work we deliberately curated biased datasets (similar to [26] in the computer vision field) to create scenarios in which shortcut learning could be demonstrated and understood and show through multiple experiments that shortcut learning based on sex and race does occur. Finally, we also provide an approach to quantify the extent of shortcut learning and identify shortcut features using interpretability.

Acknowledgments. This research was supported by the UK Engineering and Physical Sciences Research Council (EPSRC) [Grant reference number EP/Y035216/1] Centre for Doctoral Training in Data-Driven Health (DRIVE-Health) at King's College London. Data used in preparation of this article were obtained from the Alzheimer's Disease Neuroimaging Initiative (ADNI) database (adni.loni.usc.edu). As such, the investigators within the ADNI contributed to the design and implementation of ADNI and/or provided data but did not participate in analysis or writing of this report. A complete listing of ADNI investigators can be found at: http://adni.loni.usc.edu/wp-content/uploads/howtoapply/ADNIAcknowledgementList.pdf. Data were also provided in part by OASIS and the Human Connectome Project, WU-Minn Consortium (Principal Investigators: David Van Essen and Kamil Ugurbil; 1U54MH091657) funded by the 16 NIH Institutes and Centers that support the NIH Blueprint for Neuroscience Research; and by the McDonnell Center for Systems Neuroscience at Washington University.

References

1. Barocas, S., Hardt, M., Narayanan, A.: Fairness and Machine Learning: Limitations and Opportunities. MIT press, Cambridge (2023)
2. Bercea, C.I., Puyol-Antón, E., Wiestler, B., Rueckert, D., Schnabel, J.A., King, A.P.: Bias in unsupervised anomaly detection in brain mri. In: Workshop on Clinical Image-Based Procedures, pp. 122–131. Springer, Heidelberg (2023). https://doi.org/10.1007/978-3-031-45249-9_12
3. Buolamwini, J., Gebru, T.: Gender shades: intersectional accuracy disparities in commercial gender classification. In: Conference on Fairness, Accountability and Transparency, pp. 77–91. PMLR (2018)
4. Cardoso, M.J., et al.: Monai: an open-source framework for deep learning in healthcare. arXiv preprint arXiv:2211.02701 (2022)
5. Deng, J., Dong, W., Socher, R., Li, L.J., Li, K., Fei-Fei, L.: Imagenet: a large-scale hierarchical image database. In: 2009 IEEE Conference on Computer Vision and Pattern Recognition, pp. 248–255. IEEE (2009)
6. Fischl, B.: Freesurfer. Neuroimage **62**(2), 774–781 (2012)

7. Friedrich, F., et al.: Auditing and instructing text-to-image generation models on fairness. In: AI and Ethics, pp. 1–21 (2024)
8. Geirhos, R., et al.: Shortcut learning in deep neural networks. Nat. Mach. Intell. **2**(11), 665–673 (2020)
9. Gichoya, J.W., et al.: Ai recognition of patient race in medical imaging: a modelling study. Lancet Dig. Health **4**(6), e406–e414 (2022)
10. Girrbach, L., Alaniz, S., Smith, G., Akata, Z.: A large scale analysis of gender biases in text-to-image generative models. arXiv preprint arXiv:2503.23398 (2025)
11. Hammers, A., et al.: Three-dimensional maximum probability atlas of the human brain, with particular reference to the temporal lobe. Hum. Brain Mapp. **19**(4), 224–247 (2003)
12. He, K., Zhang, X., Ren, S., Sun, J.: Deep residual learning for image recognition. In: Proceedings of the IEEE Conference on Computer Vision and Pattern Recognition, pp. 770–778 (2016)
13. Irvin, J., et al.: Chexpert: a large chest radiograph dataset with uncertainty labels and expert comparison. In: Proceedings of the AAAI Conference on Artificial Intelligence, vol. 33, pp. 590–597 (2019)
14. Isensee, F., et al.: Automated brain extraction of multisequence mri using artificial neural networks. Hum. Brain Mapp. **40**(17), 4952–4964 (2019)
15. Jin, H., Xue, A., You, W., Goel, S., Wong, E.: Probabilistic stability guarantees for feature attributions. arXiv preprint arXiv:2504.13787 (2025)
16. Klingenberg, M., et al.: Higher performance for women than men in mri-based alzheimer's disease detection. Alzheimer's Res. Therapy **15**(1), 84 (2023)
17. LaMontagne, P.J., et al.: Oasis-3: longitudinal neuroimaging, clinical, and cognitive dataset for normal aging and alzheimer disease. In: Medrxiv, pp. 2019–12 (2019)
18. Liu, Z., et al.: Swin transformer: hierarchical vision transformer using shifted windows. In: Proceedings of the IEEE/CVF International Conference on Computer Vision, pp. 10012–10022 (2021)
19. Liu, Z., Luo, P., Wang, X., Tang, X.: Deep learning face attributes in the wild. In: Proceedings of the IEEE International Conference on Computer Vision, pp. 3730–3738 (2015)
20. Loshchilov, I., Hutter, F.: Decoupled weight decay regularization. arXiv preprint arXiv:1711.05101 (2017)
21. Mehrabi, N., Morstatter, F., Saxena, N., Lerman, K., Galstyan, A.: A survey on bias and fairness in machine learning. ACM Comput. Surv. (CSUR) **54**(6), 1–35 (2021)
22. Moayeri, M., Singla, S., Feizi, S.: Hard imagenet: segmentations for objects with strong spurious cues. Adv. Neural. Inf. Process. Syst. **35**, 10068–10077 (2022)
23. Oakden-Rayner, L., Dunnmon, J., Carneiro, G., Ré, C.: Hidden stratification causes clinically meaningful failures in machine learning for medical imaging. In: Proceedings of the ACM Conference on Health, Inference, and Learning, pp. 151–159 (2020)
24. Petersen, E., et al.: Feature robustness and sex differences in medical imaging: a case study in mri-based alzheimer's disease detection. In: International Conference on Medical Image Computing and Computer-Assisted Intervention, pp. 88–98. Springer, Heidelberg (2022). https://doi.org/10.1007/978-3-031-16431-6_9
25. Petersen, R.C., et al.: Alzheimer's disease neuroimaging initiative (adni) clinical characterization. Neurology **74**(3), 201–209 (2010)
26. Sagawa, S., Koh, P.W., Hashimoto, T.B., Liang, P.: Distributionally robust neural networks for group shifts: on the importance of regularization for worst-case generalization. arXiv preprint arXiv:1911.08731 (2019)

27. Selvaraju, R.R., Cogswell, M., Das, A., Vedantam, R., Parikh, D., Batra, D.: Grad-cam: visual explanations from deep networks via gradient-based localization. In: Proceedings of the IEEE International Conference on Computer Vision, pp. 618–626 (2017)
28. Shannon, C.E.: A mathematical theory of communication. Bell Syst. Techn. J. **27**(3), 379–423 (1948)
29. Stanley, E.A., Souza, R., Wilms, M., Forkert, N.D.: Where, why, and how is bias learned in medical image analysis models? A study of bias encoding within convolutional networks using synthetic data. EBioMedicine **111** (2025)
30. Stanley, E.A., Wilms, M., Mouches, P., Forkert, N.D.: Fairness-related performance and explainability effects in deep learning models for brain image analysis. J. Med. Imaging **9**(6), 061102 (2022)
31. Van Essen, D.C., et al.: The wu-minn human connectome project: an overview. Neuroimage **80**, 62–79 (2013)
32. Wang, R., Chaudhari, P., Davatzikos, C.: Bias in machine learning models can be significantly mitigated by careful training: evidence from neuroimaging studies. Proc. Natl. Acad. Sci. **120**(6), e2211613120 (2023)
33. Wang, R., Kuo, P.C., Chen, L.C., Seastedt, K.P., Gichoya, J.W., Celi, L.A.: Drop the shortcuts: image augmentation improves fairness and decreases AI detection of race and other demographics from medical images. EBioMedicine **102** (2024)
34. Winkler, J.K., et al.: Association between surgical skin markings in dermoscopic images and diagnostic performance of a deep learning convolutional neural network for melanoma recognition. JAMA Dermatol. **155**(10), 1135–1141 (2019)

Fair Dermatological Disease Diagnosis Through Auto-weighted Federated Learning and Performance-Aware Personalization

Gelei Xu[1]($\boxtimes$), Yawen Wu[2], Zhenge Jia[1], Jingtong Hu[2], and Yiyu Shi[1]

[1] University of Notre Dame, Notre Dame, IN 46556, USA
{gxu4,zjia2,yshi4}@nd.edu
[2] University of Pittsburgh, Pittsburgh, PA 15260, USA
{yawen.wu,jthu}@pitt.edu

Abstract. Dermatological diseases impact a large portion of the world's population, underscoring the importance of early diagnosis and timely intervention. To facilitate this, deep learning-based smartphone applications have been developed, leveraging federated learning to gather data while safeguarding patient privacy. However, existing federated learning frameworks are mostly designed to optimize overall performance, while the difference in diagnosis performance over demographic attributes like race, age, and gender are largely ignored. When applying federated learning to dermatological disease diagnosis, a significant diagnosis accuracy gap can occur and result in increased healthcare disparities. To obtain a fair model for all groups by using decentralized data, we propose a fairness-aware federated learning framework. Central to this framework is an adaptive weighting mechanism that dynamically emphasizes contributions from groups exhibiting suboptimal diagnostic accuracy, thereby steering the global model toward a more balanced representation. Following global training, the model is further tailored to each group through fine-tuning, promoting enhanced fairness at the local level. Experiments show that this framework improves fairness without sacrificing accuracy.

Keywords: Dermatological disease diagnosis · fairness · federated learning

1 Introduction

Dermatological diseases are one of the leading causes of non-fatal disease burden globally [5]. According to the estimation of the National Institutes of Health (NIH), one in five US citizens are at risk of developing a debilitating dermatological problem in their lifetimes [17]. Various studies have demonstrated that early diagnosis and intervention are often critical to prognosis and outcome [19]. To this end, the past decade has witnessed the rapid evolvement of smartphone apps that allow users to timely and conveniently identify issues that have emerged around their skins. Deep learning models have been widely used in these apps and show great potential for automatic diagnosing [8]. However, the lack of large-scale datasets is a major obstacle in the training of deep learning models for dermatology disease diagnosis [21]. This is primarily due to the data sharing constraints imposed by the Health Insurance Portability and Accountability Act (HIPAA) [23]. Without access to an ample dataset, achieving satisfactory results through centralized training is unfeasible.

Federated learning (FL) is a learning paradigm that enables collaboration between machine learning models while maintaining privacy by keeping data local on personal devices [18]. By leveraging FL, models can be trained on distributed mobile devices by using dermatological images locally without uploading these private images to the server [22]. However, existing FL frameworks are mostly designed to optimize the overall performance, while the local accuracy of each client is not guaranteed. For instance, the study [9] pointed out significant racial disparity for Skin Image Search, an machine learning based app that helps people identify skin conditions. It reports 70% accuracy for the whole dataset, but only 17% for dark skins. The heavy skin-type bias is a result of several factors, including a lack of medical professionals from marginalized communities, insufficient information about those communities, and socioeconomic barriers to data collection and research. What is worse, inadequate data could also lead to misdiagnosing people with brown and black skin, exacerbating healthcare disparities. Several works aim to improve fairness via FL [7,11,12,14]. However, the fairness level tends to fluctuate during the training process, which necessitates extensive hyperparameter tuning, particularly when dealing with highly imbalanced dermatological disease datasets. Currently, no existing approach has successfully tackled the fairness problem in this domain.

To address this problem, we develop a federated framework that proactively identifies under-represented clients and enhances model performance across these clients to promote fairness. Within this framework, a dynamic weighting mechanism is introduced to adaptively amplify the influence of clients with elevated loss, encouraging the global model to shift its representation toward underserved data distributions. After global optimization, each client further refines its model locally, building upon the shared initialization. To mitigate persistent disparities, clients exhibiting lagging performance are adaptively guided through additional optimization. This collaborative process not only improves individual client accuracy but also harmonizes performance across the population. We evaluate our framework on two dermatological disease datasets, demonstrating substantial fairness improvement without compromising overall diagnostic accuracy. Notably, while fairness is often considered a trade-off with diagnosis performance, our results demonstrate that our proposed method can achieve fairness without sacrificing overall classification performance.

2 Background and Related Work

Federated Learning for Dermatological Disease Diagnosis. FL enables knowledge sharing among numerous decentralized edge devices while keeping raw data local [10]. In a typical FL framework *FedAvg* [13,24], there are a central server and multiple distributed clients. In *FedAvg*, learning is performed round-by-round. In each round, the model parameters θ are updated by using the learned parameters θ_c^{t+1} from each client following $\theta^{t+1} \leftarrow \sum_{c \in C^t} \frac{|D_c|}{\sum_{i \in C^t} |D_i|} \theta_c^{t+1}$, where C^t represents the active clients in round t and D_c represents the local data of the client c. This learning process is repeated until the model converges.

FL has been utilized in dermatological disease diagnosis in [1,6,22]. However, these studies have focused on achieving high classification accuracy, neglecting the crucial issue of fairness, which can result in healthcare disparities. As such, there is a pressing need for an FL approach that tackles the fairness problem in dermatological disease diagnosis.

Fairness in Federated Learning. Ensuring fairness among clients in FL is crucial, as it impacts their willingness to participate, which further has a significant influence on the performance of the learned model. To achieve fairness in FL, [14] proposed a framework that minimizes the loss of the most challenging target distribution by creating a blend of the client distributions. [11] proposed *Ditto* using multi-task and federated learning to create personalized classifiers for clients to increase the accuracy of low-performance clients. [7,12] proposed algorithms for a more uniform accuracy distribution by tuning the fairness-related hyperparameters. However, most of the methods rely on several hyperparameters that demand meticulous tuning and may lead to biased models if these hyperparameters are not adequately tuned. Therefore, a method to automatically recognize clients associated with under-represented demographic groups and improve the model's performance on these clients is needed.

3 Methodology

3.1 Overview of *FedAutoFair*

Figure 1 illustrates an overview of the proposed framework. The process begins with federated optimization, where clients iteratively update local models and share parameters along with aggregated training loss metrics. A centralized auto-weighting mechanism adjusts each client's contribution based on relative loss, with the scaling factor dynamically inferred from the client distribution. The server then aggregates updates through weighted averaging to obtain the global model. After convergence, clients further adapt the global model through local refinement. To reduce disparity, clients with lower performance are selectively prioritized for additional optimization. This process yields locally adapted models with improved diagnostic accuracy and enhanced fairness.

3.2 Fairness-Aware Automatic Weighting

To automatically identify clients from under-represented demographic groups and enhance the model's performance for them, we propose a seemingly simple, yet extremely effective method for fair dermatological disease diagnosis. To

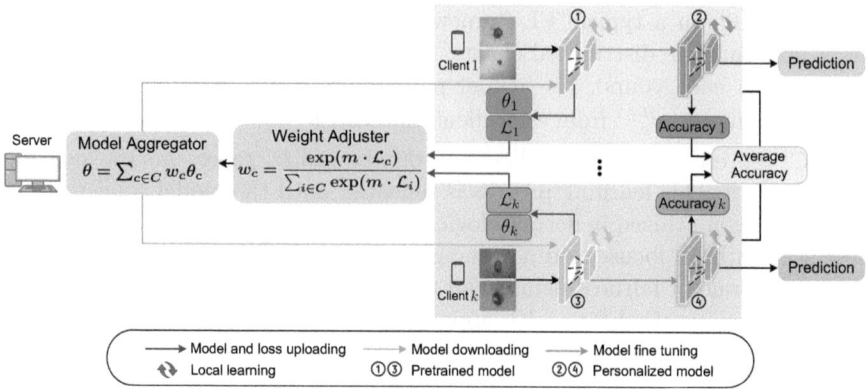

Fig. 1. The *FedAutoFair* framework for fair dermatological disease diagnosis. An adaptive weighting module dynamically adjusts client contributions based on training loss across communication rounds. Upon global model convergence, clients undergo performance-aware refinement to derive personalized models.

reduce the performance gap among clients, our approach assigns more weight to clients with poorer performance such that their performance can be improved and fairness is enhanced. In this method, for each client, we measure the model's performance by using the loss value of the model on the local data. The average loss $\mathcal{L}_c$ for client c is calculated as follows:

$$\mathcal{L}_c = \frac{1}{N_c} \sum_{n=1}^{N_c} l_{ce}(p_n, t_n) = -\frac{1}{N_c} \sum_{n=1}^{N_c} \sum_{i=1}^{M} t_n^{(i)} \log(p_n^{(i)}), \quad (1)$$

where N_c represents the number of data samples in client c, and M is the total number of classes. The function $l_{ce}(p_n, t_n)$ represents the cross-entropy loss function. $p_n^{(i)}$ is the predicted probability of sampling n belong to class i. t_n is the groud truth of data sample n.

Based on the loss function of each client $\mathcal{L}_c$, the auto-weighting function is constructed as shown below:

$$w_c = \frac{\exp(m \cdot \mathcal{L}_c)}{\sum_{i \in C} \exp(m \cdot \mathcal{L}_i)}, \quad (2)$$

where C is the set of all clients, and m is a parameter that is automatically adjusted to achieve fairness.

The auto-weighting function Eq. (2) is designed based on two ideas. First, the value of parameter m is adaptively adjusted during the learning process based on the loss values of clients. m will increase if the difference between the minimum and maximum loss value is large, indicating that some client has poor performance and needs more attention in model aggregation. Second, an exponential function wraps the loss value to enhance the differences. Larger loss values will be amplified, while smaller loss values will be reduced. As a result, an under-performed client with a larger loss value will have a larger weight during model aggregation. By using the above weights, model aggregation is performed and the global model is updated by $\theta = \sum_{c \in C} w_c \theta_c$. Subsequently, the updated global model is distributed to clients to initiate the next round of training.

Algorithm 1. *FedAutoFair* with automatic weighting.

Input: The set of all clients C, round number T, local epoch number E, batch size B, model parameter θ, hyperparameter $\mathcal{Q}, \mathcal{M}$
Output: Trained fair aggregated model θ
$m \leftarrow 1$
for each round $t = 1, 2, 3...T$ **do**
  $L \leftarrow \emptyset$
  for each client $c \in C$ in parallel **do**
    $\theta_c \leftarrow \theta$ ▷ *distribute global model*
    $\theta_c, \mathcal{L}_c \leftarrow \text{Update}(c, \theta_c)$
    L append $\mathcal{L}_c$
  end for
  if $L_{max} > \mathcal{Q} \times L_{min}$ and $m \leq \mathcal{M}$ **then**
    $m \leftarrow m + 1$
  end if
  for each client $c \in C$ **do**
    $w_c \leftarrow w_c = \frac{\exp(m \cdot \mathcal{L}_c)}{\sum_{i \in C} \exp(m \cdot \mathcal{L}_i)}$ ▷ *auto weighting*
  end for
  $\theta \leftarrow \sum_{c \in C} w_c \theta_c$ ▷ *aggregate models*
end for
Update(k, θ_k):   ▷ *run on client k*
$\mathcal{B} \leftarrow$ (split dataset of k into batches of size B)
for each local epoch i from 1 to E **do**
  for batch $b \in \mathcal{B}$ **do**
    $\theta_k \leftarrow \theta_k - \eta \nabla \mathcal{L}(\theta_k; b)$
  end for
end for
return $\theta_k, \mathcal{L}_k$ to server

With the auto-weighting function Eq. (2), the learning process is described in Algorithm 1. The client performs training locally and updates the model parameters using the **Update** method in each round. To assign weights, m in the range of $(0, \mathcal{M})$ is utilized. If the maximum client loss becomes Q times larger than the minimum loss, m will be increased by 1. Finally, the models are aggregated on the server by using the auto-weighting function.

3.3 Fairness-Aware Adaptation

To further enhance fairness, we introduce a post-aggregation adaptation phase tailored to each client. While model refinement is typically employed to boost local performance, our findings suggest it can also serve as an effective lever for fairness. Given the inherent heterogeneity in dermatological data—such as variations in skin type and condition—a single global model may struggle to generalize uniformly across clients. In particular, clients with harder-to-classify samples benefit from individualized adaptation that builds on the shared model. To mitigate persistent disparities, clients exhibiting the lowest diagnostic accuracy are selectively prioritized for additional refinement, ensuring underperforming groups receive targeted improvements. This strategy leads to more balanced representation learning in both equity and overall diagnostic quality.

4 Experiments and Results

Dataset. The effectiveness of our proposed method was assessed using two datasets: the Fitzpatrick 17k dataset [4] and the ISIC 2019 Challenge dataset [2,20]. The Fitzpatrick 17k dataset consists of 16,577 images that display nine different skin conditions, as defined by the Fitzpatrick scoring system. The dataset includes images of six different skin types, with 2,944, 4,807, 3,306, 2,781, 1,531, and 634 images representing skin types 1 through 6, respectively [4]. The ISIC 2019 dataset contains 24,894 images in 9 diagnostic categories. We use age to group the data and there are 684, 2,264, 6,482, 6,695, 5,991, and 2,778 images of age types 1 through 6, respectively. The highly skewed dataset poses a challenging but realistic setting for our methods as groups with few data points may not be fully trained, potentially causing fairness issues.

Preprocessing Details. The dataset is randomly split into a training set of 60% data, a validation set of 20% data, and a test set of 20% data, respectively. To accommodate the limited data of clients, We divided the training set into a public dataset, consisting of one-sixth of the data accessible to all clients, and private datasets containing clients' local images. Data augmentation techniques including horizontal and vertical flipping, rotation, scaling, and autoaugment are used to augment the private data [3]. All input images were resized to 128×128, and resampling was used to ensure class balance.

Federated Setting and Training Details. 127 and 209 clients are used in Fitzpatrick 17k and ISIC 2019 datasets respectively and each client consists of images of a single fair attribute. FL is performed for 100 rounds. In each round, each client trained for five local epochs. The VGG-11 backbone was used with an Adam optimizer, a batch size of 128, and an initial learning rate of 1e–4 with a cosine decay. The hyper-parameter $\mathcal{M}$ is set to 3 and $\mathcal{Q}$ is set to $\frac{3}{2}$ in Algorithm 1. In the fine-tuning stage, each client was trained for 50 rounds. In each round, we identified the clients whose validation accuracy fell in the bottom 30% of all clients and trained them for an additional five epochs to reduce classification disparities. All experiments were conducted on an Nvidia V100 GPU.

Baselines. We compared our proposed method with five baselines. *FedAvg* [13] is a standard FL aggregation method that assigns weights to clients based on the number of samples they provide. *q-FFL* [12] achieves fairness in FL by manually tuned hyper-parameters. *Ditto* [11] utilizes the squared distance between local and global models to train an extra local model for fairness. *FedFa* [7] achieves fairness by a weight selection technique. We further create three variants of *FedAvg*: *FedEqual*, *FedLoss*, and *FedExp*, with different weights for model aggregation. *FedEqual* sets the weights of all clients equally. *FedLoss* uses the normalized loss values as the weights, while *FedExp* uses the normalized exponential loss values as the weights.

Fairness Metrics. Two metrics are used in this paper to evaluate fairness.

Variance of Accuracy: Since the goal of fairness is to achieve accuracy parity by measuring the uniformity across FL devices [12], the degree of uniformity between different groups can be measured by the variance of accuracy [16]. For this metric, a smaller value means better fairness.

Gap and Worst: We also quantify fairness by extending the *gap* and *worst* metric in [15] to multi-class classification. Specifically, we utilize the *gap* and *worst* metrics to evaluate each group s. Here, $Acc(s)$ represents the accuracy of group s, while $Acc(\neg s)$ represents the accuracy of all other groups except for s. $Acc^{gap}(s)$ and $Acc^{worst}(s)$ are defined as follows.

$$Acc^{gap}(s) = |Acc(s) - Acc(\neg s)|, \qquad (3)$$

$$Acc^{worst}(s) = \min(Acc(s), Acc(\neg s)). \qquad (4)$$

Intuitively, a fair model minimizes the accuracy gap and maximizes the worst accuracy for each group to ensure both fairness and accuracy. Therefore, group s with lower $Acc^{gap}(s)$ and higher $Acc^{worst}(s)$ will contribute more in fairness.

Table 1. The results of proposed *FedAutoFair* and baselines on two datasets, with the variance scale in 10^{-3}.

Method	Fitzpatrick 17k				ISIC 2019			
	Precision	Recall	F1	Variance↓	Precision	Recall	F1	Variance↓
FedAvg [13]	0.582	0.622	0.598	1.031	0.695	0.724	0.698	1.332
q-FFL [12]	0.609	0.636	0.615	0.763	0.709	0.728	0.719	1.244
Ditto [11]	0.613	0.627	0.620	0.570	0.689	0.725	0.710	1.033
FedFa [7]	0.604	0.620	0.613	0.801	0.702	0.720	0.712	1.123
FedEqual	0.591	0.638	0.609	0.882	0.697	0.713	0.694	1.607
FedLoss	0.608	0.639	0.618	0.291	0.683	0.722	0.692	1.228
FedExp	0.593	0.643	0.622	0.535	0.709	0.730	0.719	1.165
FedAutoFair (ours)	0.623	0.649	0.628	0.123	0.715	0.730	0.718	0.265

Main Results. Table 1 compares the precision, recall, F1, and variance of different methods on two datasets. Our *FedAutoFair* achieves the best fairness and best performance on Fitzpatrick 17k dataset. Notably, while fairness is often considered a trade-off with classification performance, our results demonstrate that our proposed method can achieve fairness without sacrificing overall classification performance. *FedAutoFair* has the smallest variance, over 4× smaller than the *Ditto* method, and also attains the highest precision, recall, and F1.

Similarly, on the ISIC 2019 dataset, our *FedAutoFair* effectively decreases the variance while maintaining high performance. The other methods, on the other hand, do not significantly reduce the variance.

Group-Wise Fairness Analysis. To provide a comprehensive evaluation of the fairness of both baseline models and *FedAutoFair*, we present a radar plot in Fig. 2 showing the accuracy of different demographic groups for Fitzpatrick 17k and ISIC 2019 datasets. As *FedEqual*, *FedLoss* and *FedExp* use similar fairness techniques and Table 1 demonstrates that *FedExp* achieved the best results, we'll solely report *FedExp*'s results hereafter. The figure illustrates that in both datasets, our proposed method achieves the highest accuracy for most fairness groups with a particular improvement in the challenging-to-classify (group 5, 6 in Fitzpatrick 17k and group 1, 6 in ISIC 2019) resulting in accuracy scores that resemble a hexagon. On the other hand, the accuracy of *FedAvg* is highly imbalanced, while the effectiveness of *q-FFL*, *Ditto*, *FedFa*, and *FedExp* is unstable.

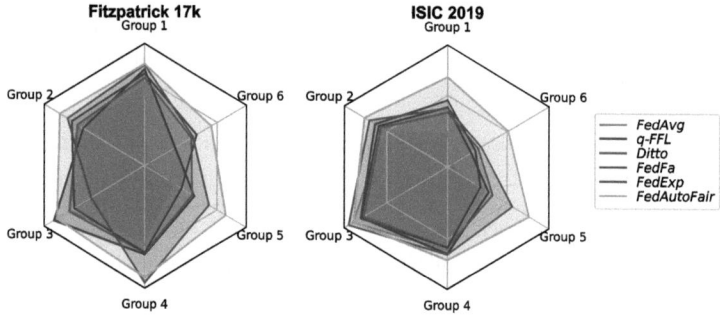

Fig. 2. The accuracy with respect to different demographic groups on Fitzpatrick 17k and ISIC 2019 datasets (best viewed in color).

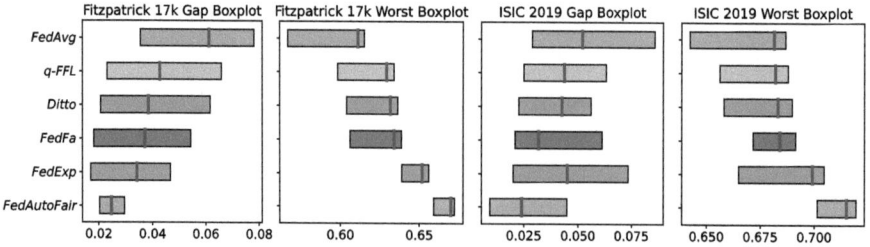

Fig. 3. The *gap* and *worst* metrics. *gap* is the accuracy difference between group s and $\neg s$, *worst* is the minimum accuracy of s and $\neg s$. The box shows the data range, with the median represented by a red line (best viewed in color).

Fairness Metric Evaluation. We assess fairness using two complementary metrics: *gap* (disparity across groups) and *worst* (performance of the least-served group), aiming to minimize the former and maximize the latter, while maintaining consistency across groups. As shown in Fig. 3, *FedAvg*, which lacks fairness mechanisms, exhibits the widest range and worst median across both metrics. While *q-FFL*, *Ditto*, *FedFa*, and *FedExp* offer moderate improvements, their performance remains inconsistent across datasets. In contrast, *FedAutoFair* consistently achieves the most favorable fairness scores—yielding both lower *gap* and higher *worst*—demonstrating superior robustness and equity. These findings highlight our method's ability to reduce inter-group disparities while elevating the performance of under-represented groups.

5 Conclusion

In this paper, we present a framework that employs federated learning to address fairness issues in dermatological disease diagnosis. Our approach ensures fairness by assigning higher weights to clients with greater losses through our auto-weighted method. Each client fine-tunes its local model based on the global model, resulting in superior fairness and diagnosis performance. Experiments on two datasets demonstrate the efficacy of our proposed methodology in improving fairness without sacrificing any accuracy in dermatological disease diagnosis.

Acknowledgments.. This project is supported in part by NIH Grant R01EB033387.

Disclosure of Interests. The authors declare no competing interests.

References

1. Agbley, B.L.Y., et al.: Multimodal melanoma detection with federated learning. In: 2021 18th International Computer Conference on Wavelet Active Media Technology and Information Processing (ICCWAMTIP), pp. 238–244. IEEE (2021)
2. Combalia, M., et al.: Bcn20000: dermoscopic lesions in the wild. arXiv preprint arXiv:1908.02288 (2019)
3. Cubuk, E.D., Zoph, B., Mane, D., Vasudevan, V., Le, Q.V.: Autoaugment: learning augmentation policies from data. arXiv preprint arXiv:1805.09501 (2018)
4. Groh, M., et al.: Evaluating deep neural networks trained on clinical images in dermatology with the fitzpatrick 17k dataset. In: Proceedings of the IEEE/CVF Conference on Computer Vision and Pattern Recognition, pp. 1820–1828 (2021)
5. Hollestein, L.M., Nijsten, T.: An insight into the global burden of skin diseases. J. Invest. Dermatol. **134**(6), 1499–1501 (2014)
6. Hossen, M.N., Panneerselvam, V., Koundal, D., Ahmed, K., Bui, F.M., Ibrahim, S.M.: Federated machine learning for detection of skin diseases and enhancement of internet of medical things (iomt) security. IEEE J. Biomed. Health Inf. (2022)
7. Huang, W., Li, T., Wang, D., Du, S., Zhang, J., Huang, T.: Fairness and accuracy in horizontal federated learning. Inf. Sci. **589**, 170–185 (2022)

8. Jones, M., Morris, J., Deruyter, F.: Mobile healthcare and people with disabilities: current state and future needs. Int. J. Environ. Res. Public Health **15**(3), 515 (2018)
9. Kamulegeya, L.H., et al.: Using artificial intelligence on dermatology conditions in Uganda: a case for diversity in training data sets for machine learning. BioRxiv p. 826057 (2019)
10. Konečnỳ, J., McMahan, B., Ramage, D.: Federated optimization: distributed optimization beyond the datacenter. arXiv preprint arXiv:1511.03575 (2015)
11. Li, T., Hu, S., Beirami, A., Smith, V.: Ditto: fair and robust federated learning through personalization. In: International Conference on Machine Learning, pp. 6357–6368. PMLR (2021)
12. Li, T., Sanjabi, M., Beirami, A., Smith, V.: Fair resource allocation in federated learning. arXiv preprint arXiv:1905.10497 (2019)
13. McMahan, B., Moore, E., Ramage, D., Hampson, S., Arcas, B.A.: Communication-efficient learning of deep networks from decentralized data. In: Artificial Intelligence and Statistics, pp. 1273–1282. PMLR (2017)
14. Mohri, M., Sivek, G., Suresh, A.T.: Agnostic federated learning. In: International Conference on Machine Learning, pp. 4615–4625. PMLR (2019)
15. Ramapuram, J., Busbridge, D., Webb, R.: Evaluating the fairness of fine-tuning strategies in self-supervised learning. arXiv preprint arXiv:2110.00538 (2021)
16. Shi, Y., Yu, H., Leung, C.: A survey of fairness-aware federated learning. arXiv preprint arXiv:2111.01872 (2021)
17. Stern, R.S.: Prevalence of a history of skin cancer in 2007: results of an incidence-based model. Arch. Dermatol. **146**(3), 279–282 (2010)
18. Tan, A.Z., Yu, H., Cui, L., Yang, Q.: Towards personalized federated learning. IEEE Trans. Neural Netw. Learn. Syst. (2022)
19. Thomsen, K., Iversen, L., Titlestad, T.L., Winther, O.: Systematic review of machine learning for diagnosis and prognosis in dermatology. J. Dermatol. Treat. **31**(5), 496–510 (2020)
20. Tschandl, P., Rosendahl, C., Kittler, H.: The ham10000 dataset, a large collection of multi-source dermatoscopic images of common pigmented skin lesions. Sci. Data **5**(1), 1–9 (2018)
21. Wu, Y., Wang, Z., Shi, Y., Hu, J.: Enabling on-device cnn training by self-supervised instance filtering and error map pruning. IEEE Trans. Comput. Aided Des. Integr. Circuits Syst. **39**(11), 3445–3457 (2020)
22. Wu, Y., et al.: Federated contrastive learning for dermatological disease diagnosis via on-device learning. In: 2021 IEEE/ACM International Conference On Computer Aided Design, pp. 1–7. IEEE (2021)
23. Wu, Y., Zeng, D., Wang, Z., Shi, Y., Hu, J.: Federated contrastive learning for volumetric medical image segmentation. In: de Bruijne, M., et al. (eds.) MICCAI 2021. LNCS, vol. 12903, pp. 367–377. Springer, Cham (2021). https://doi.org/10.1007/978-3-030-87199-4_35
24. Zhu, W., Luo, J.: Federated medical image analysis with virtual sample synthesis. In: Medical Image Computing and Computer Assisted Intervention–MICCAI 2022: 25th International Conference, Singapore, 18–22 September 2022, Proceedings, Part III, pp. 728–738. Springer, Heidelberg (2022). https://doi.org/10.1007/978-3-031-16437-8_70

Assessing Annotator and Clinician Biases in an Open-Source-Based Tool Used to Generate Head CT Segmentations for Deep Learning Training

Artur Paulo[1(✉)], Pedro Vinicius Silva[1], Tayran Mila Mendes Olegario[1], Paula Bresciani de Andrade[1], Klaus Schumacher[1], Rafael Maffei Loureiro[1], Joselisa Peres Queiroz de Paiva[1], Raissa Souza[2], and Bruna Garbes Gonçalves Pinto[1]

[1] Hospital Israelita Albert Einstein, São Paulo, Brazil
artur.pauloj@gmail.com
[2] Department of Radiology, Cumming School of Medicine, University of Calgary, Calgary, Canada

Abstract. Machine learning (ML) has shown strong performance in medical imaging analysis. However, most models rely on large, high-quality annotated datasets, which are time-consuming, costly, and require domain expertise. Recent studies have explored ML-based methods for generating annotations. While promising, these approaches demand considerable computational resources and rely on the initial model's performance, posing challenges in low- and middle-income countries with limited infrastructure. To address these limitations, we developed a novel, open-source-based tool for efficient and transparent medical image annotation using a human-in-the-loop (*i.e.,* three bioscience students and five neuroradiologists) approach. Although the annotations generated were suitable to train downstream ML models, it is important to evaluate if and how the annotators and clinicians can introduce biases into the datasets when using our tool. Thus, in this work, we analyzed interobserver agreement and used mixed-effect models to identify potential biases in the annotation workflow. Our findings showed that interobserver agreement for intracranial volume segmentations ranged from substantial to nearly perfect, indicating strong consistency among neuroradiologists. Regression analysis revealed that neither annotator identity nor annotation time significantly affected acceptance rates, suggesting that structured training and feedback may have reduced variability. These aspects are essential to build trust and incentivize the community to utilize our tool, which represents a practical and equitable solution for medical imaging research in low-resource settings.

Keywords: Annotator bias · Clinician bias · Open-source tool · Human-in-the-loop · Fairness · Low-resource settings

---

R. Souza and G. G. Pinto—Shared last authorship.

1 Introduction

The growing reliance on machine learning (ML) in medical imaging analysis has raised concerns about algorithmic fairness and global representation, as most training datasets are created in high-income regions with limited inclusion of diverse populations. This imbalance risks propagating biases into diagnostic tools, which can decrease their reliability and safety when deployed in underrepresented regions. Computer-aided diagnosis (CAD) systems for medical imaging are often based on supervised learning paradigms, demanding extensive, high-quality datasets with expert-level annotations for optimal performance [17]. While ML-generated pseudo-labels offer a partial solution [11,17], their effectiveness is closely tied to the initial accuracy of the models. Consequently, such solutions may be infeasible in low- and middle-income countries, where access to high-performance computing and domain expertise remains limited.

To address these limitations, we developed an open-source-based, human-in-the-loop annotation framework. Our platform integrates containerized infrastructure, open-source imaging tools, and a web-based dashboard. Following structured training, bioscience students performed initial segmentations of intracranial and lateral ventricle volumes in head CT scans. These annotations were then validated by board-certified neuroradiologists through a transparent, auditable workflow. This system was used by Pinto et al. (2024) to generate labels for a deep learning segmentation model [14], demonstrating its utility for real-world tasks. However, it is essential to evaluate if and how the annotators and clinicians involved in the workflow could introduce biases to the generated brain segmentation annotations. These biases may arise from differences in experience level, years of study in the case of bioscience students, task interpretation, and inherent human subjectivity when judging where tissue boundaries are [15].

This work presents an analysis of interobserver agreement to identify such potential biases and flaws that could impair the reliability of the annotations. Furthermore, we describe the annotation flow audits for reproducibility and assess potential sources of bias in the annotation and validation processes. These aspects are crucial to building trust and incentivizing the community to utilize our open-source-based tool. This work demonstrates an example of the tool's utility in generating reliable annotations for head CT scans employing a human-in-the-loop approach. Since it enables collaboration between experts and students, it promotes equitable participation in data curation. Most importantly, it contributes to the scientific community by presenting a practical solution for medical imaging research in low-resource settings, paving the way to include regions historically excluded from dataset curation initiatives.

2 Materials and Methods

2.1 Data

We retrospectively collected 481 volumetric head CT scans from the Hospital Israelita Albert Einstein in São Paulo, Brazil. The requirement for individual consent was waived by the local Ethics Committee (approval number: 52257521.8.0000.0071).

A subsample comprising 80 head CT scans was selected and segmented by three annotators (A1, A2, A3), as illustrated in Fig. 1. Each annotator segmented a different number of scans: A1 segmented 24 images, A2 segmented 29, and A3 segmented 27. The resulting segmentation masks were subsequently reviewed by five neuroradiologists, who were blinded to the identity of the annotators, resulting in a total of 400 observations.

2.2 Annotator Selection

The project was promoted with support from academic offices and through online student groups at universities. The selection of annotators was based on knowledge of neuroanatomy and, preferably, prior experience with the 3D Slicer software. Candidates had a background in bioscience and participated in a training session. They also performed a segmentation challenge using CT scans from the CQ500 dataset [3]. From 30 applicants, ten were selected based on their experience and the quality of their segmentations. During the project, seven students withdrew due to other professional opportunities, and the remaining three completed their activities, which were sufficient to meet the project's demands. All selected students received equal monthly stipends.

2.3 Annotation Pipeline

The dataset was preprocessed following the protocol in [14], including the selection of non-contrast scans with soft tissue kernels, exclusion of scans without thin-section series (≤ 1.3 mm pixel spacing), and removal of images with major artifacts or missing metadata (age, sex). Selected scans were converted to NIfTI format and stored in a cloud-based storage system.

A heuristic pre-segmentation of the intracranial region–based on thresholding, erosion, connected components, and dilation–was implemented using SimpleITK [10]. Annotators accessed the images through a customized 3D Slicer interface [5], running in Docker containers on cloud virtual machines secured via VPN. This setup enabled fast loading of scans and pre-segmentations. Final masks were submitted using a "send image" button, and a comment field allowed communication with neuroradiologists. Auto-saving was enabled to prevent data loss.

Each annotator worked in an isolated Docker environment connected to a PostgreSQL database, which handled segmentation flow and communication between annotators and validators via Python scripts. Real-time progress tracking, automated backups, and structured logging ensured data integrity, auditability, and reproducibility.

2.4 Validation Pipeline

The validation was carried out by five board-certified radiologists with a median of 6 years (range: 4–9 years) of experience, all of whom were blinded to the iden-

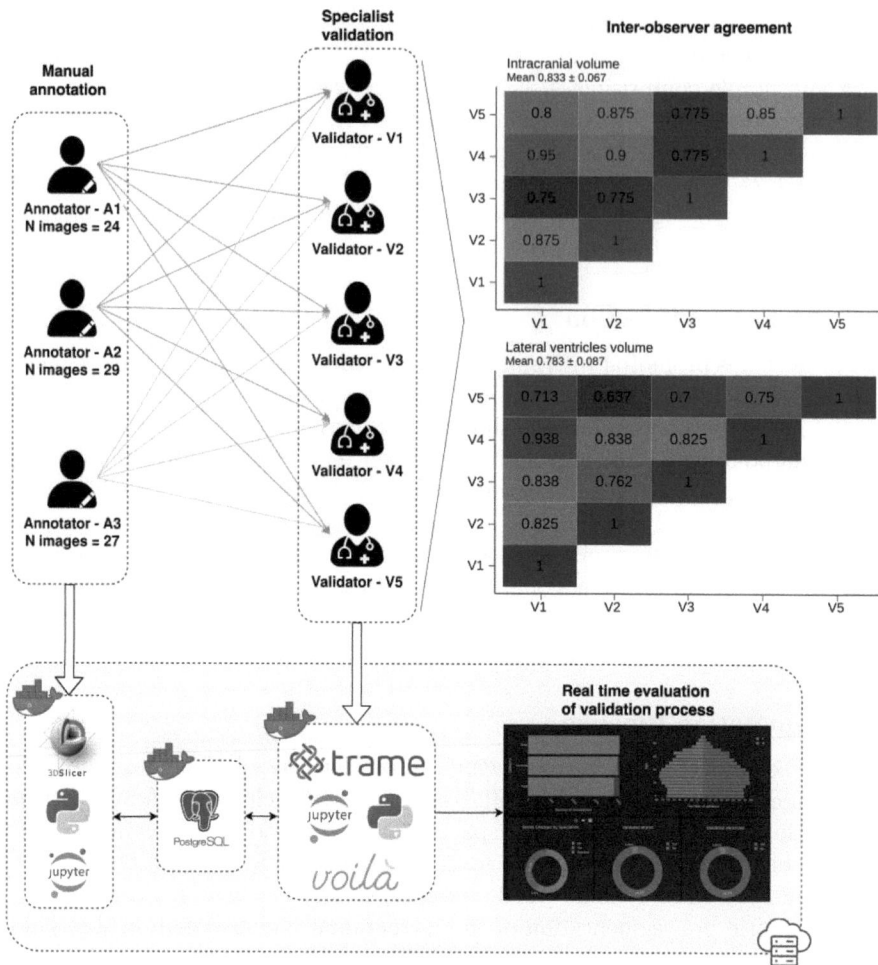

Fig. 1. Scheme showing the annotation and validation process to perform the inter-observer agreement analysis. A similar process was implemented in the infrastructure architecture to allow the annotation of the whole dataset. The heatmaps display the pairwise acceptance rates between five different validators for both segmented structures. The mean acceptance rate with standard deviation is also provided for each volume measurement

tity of the annotators. Validators used a dashboard built with the Trame framework [1] and the Voilá/Jupyter environment [2] to access a validation panel. This panel allowed neuroradiologists to visually inspect cranial CT images and the segmentations generated by annotators. They could scroll through image slices using a slider, approve or reject images and intracranial/ventricular masks via checkboxes, view annotator comments, and provide written feedback. The feedback mechanism between validators and annotators introduces a form of accountable oversight, reinforcing transparency throughout the annotation process. Rejected masks were returned to annotators with comments for refinement, remaining in a PostgreSQL-managed queue until they were approved. The system tracked the number of refinement rounds per mask. For the agreement study, only one refinement round was allowed to assess inter-validator acceptance.

2.5 Monitoring of Results

Another tab of the dashboard displayed real-time metrics on the number of annotations, stratified by segmentation type, patient sex, and age. It also tracked rejected images, such as contrast-enhanced CTs not previously flagged due to incomplete or inaccurate DICOM metadata.

Although not shown on the dashboard, timestamps and task durations for annotations and validations were stored in the database. These data enabled analyses of annotator effort and average annotation volume, contributing to project monitoring and accountability.

2.6 Open-Source-Based Tool Implementation

We developed a containerized annotation pipeline using Docker and PostgreSQL, integrating 3D Slicer for segmentation and Trame for visualization. PostgreSQL served as the central database to track scan assignments, annotation status, and quality control metrics, as shown at the bottom of Fig. 1. The system was designed to be annotation-agnostic and can be readily adapted for other tasks such as bounding-box or slice-level annotations and image modalities (*e.g.*, MRI, or X-ray).

As previously described, annotators accessed the system via a local web interface that enabled seamless image loading, segmentation saving, submission for expert review, and receipt of expert feedback. The dashboard also provided real-time metrics, such as the number of images segmented by each annotator and the number of images validated by each radiologist.

2.7 Interobserver Agreement and Bias Analysis

To detect potential validator biases, such as an individual neuroradiologist being overly rigorous or an annotator consistently producing poor-quality segmentations, we employed Cohen's Kappa to quantify the pairwise agreement between neuroradiologists for each anatomical structure. Cohen's Kappa interpretation followed Landis and Koch's scale: slight (0.00–0.20), fair (0.21–0.40), moderate (0.41–0.60), substantial (0.61–0.80), and almost perfect (0.81–1.00). We did not calculate the global agreement of the five validators in this work, as this analysis has already been presented in another work by our group [14].

To evaluate the consistency of annotator performance and the stringency of each validator, we computed the acceptance rate, corresponding to the ratio of accepted segmentations over the total submissions per annotator-validator pair.

To assess biases related to annotator identity and annotation time on acceptance rates for a given mask, we implemented a generalized mixed-effects model. We chose to focus this analysis on the segmentation of the ventricles because it relied exclusively on manual annotation (without automatic pre-segmentation). Given the crossed experimental design (Fig. 1), both image and validator were treated as random intercepts, while annotator and annotation time were included as fixed effects. A two-step sensitivity analysis was conducted on the results of the mixed-effect model. First, influential observations at the image level were identified using Cook's distance, with values exceeding $4/80 = 0.05$ considered influential. Then, these points were removed from the model to assess their effect on the regression coefficients and statistical significance. Second, we summarized the mean annotation time for each annotator and excluded observations with durations exceeding two standard deviations to remove instances where the 3D Slicer environment was left running.

3 Results

3.1 Validator Biases Analyses Based on Interobserver Agreement

Interobserver agreement among neuroradiologists for intracranial volume segmentations ranged from substantial ($\kappa = 0.775$) to almost perfect ($\kappa = 0.875$), indicating strong concordance among validators. For the lateral ventricles, the magnitude of agreement ranged from substantial ($\kappa = 0.637$) to almost perfect ($\kappa = 0.938$). In both cases, the results suggest that the validator's experience level is not acting as a strong source of bias.

Figure 1 shows a higher mean agreement for intracranial volume segmentation. This is expected given that the lateral ventricles are more nuanced and heterogeneous structures, which makes their segmentation inherently more complex. Nevertheless, the mean agreement was classified as almost perfect for intracranial volume and substantial for the lateral ventricles. These findings suggest that even with more complex brain structures, the validator's experience level did not bias their evaluations.

3.2 Annotator Biases Assessment

The sensitivity analysis of the mixed-effect model was conducted by removing 3 influential points and 25 outliers. The model assessed whether annotation time and annotator identity were associated with the likelihood of segmentation acceptance. Annotator 1 was used as the reference category for comparisons. The results showed no statistically significant effect of annotation time (p = 0.43), Annotator 2 (vs. Annotator 1; p = 0.22), Annotator 3 (vs. Annotator 1; p = 0.20), or their interactions with annotation time (Annotator 2 × time: p = 0.63; Annotator 3 × time: p = 0.77). These results remained consistent after both sensitivity steps, suggesting no meaningful influence of annotator identity or time on ventricular segmentation outcomes.

The acceptance rate between validators and annotators is shown in Fig. 2. Annotator 3 (A3) had the lowest acceptance rate in nine out of ten validator-structure combinations. We believe this may be due to lower availability, as A3 consistently produced fewer segmentations per week when compared to the other annotators when annotating the full dataset. Although no production targets were set, shared weekly averages might have led A3 to prioritize quantity over quality in an effort to keep up. However, it is essential to note that we were unable to confirm this hypothesis. Importantly, these differences were not statistically significant in the regression analysis, aligning with previous findings that annotator identity and annotation time are not sources of bias.

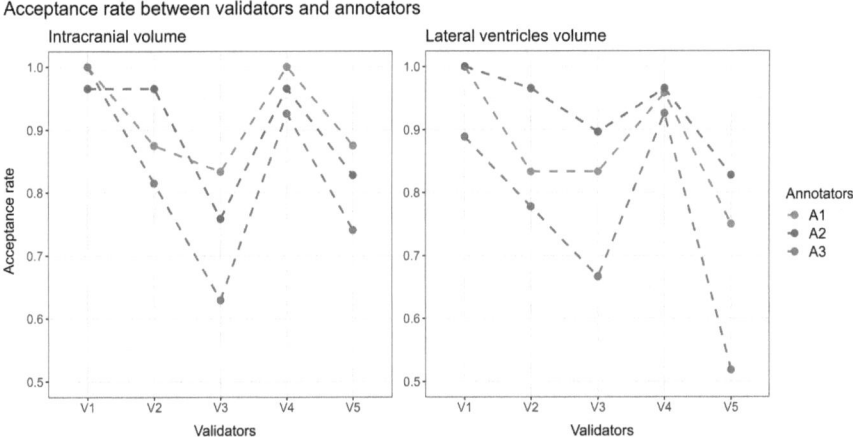

Fig. 2. Acceptance rate between validators and annotators for intracranial volume and lateral ventricles volume.

4 Discussion

This work presents an analysis of potential sources of biases in our open-source-based tool that uses a human-in-the-loop approach. It is worth mentioning that our tool was created to address medical imaging challenges in low-resource settings. Our results demonstrate high interobserver agreement among neuroradiologists, especially for intracranial volume segmentations. These results are consistent with the global agreement analysis performed with these same data, published in another study by our group [14], which showed a Gwet's AC1 statistic agreement of 0.78 [0.70;0.87] for the intracranial region and 0.71 [0.62;0.80] for the lateral ventricles. According to the Landis' criterion, these values are considered substantial for both regions. The results of this paper also highlight the high quality of annotations produced through a semi-automated pipeline involving trained bioscience students and expert reviewers. These findings support previous research indicating that hybrid human-machine systems can match the annotation quality of fully expert-led processes [8]. Most importantly, without introducing biases related to validator experience level, annotator identity, and annotation time.

Importantly, our tool operates on commodity hardware without GPU acceleration, highlighting its accessibility in environments with limited computational resources. Unlike ML-based pseudo-labeling, which requires pre-trained models and significant computing power [11], our framework minimizes technical dependencies while preserving annotation integrity. Its containerized design ensures reproducibility, portability, and adaptability to various imaging tasks, including MRI, X-rays, and bounding-box or slice-level annotations.

A distinctive feature of our platform is the tracking of interobserver agreement metrics across annotators and validators, allowing targeted feedback to enhance accuracy and consistency, which are essential for skill development in annotation tasks. Moreover, this data can help identify biases among individual validators [6,12]. By aggregating consensus from multiple experts, we mitigate the risk of introducing systematic biases from any single validator, which is crucial for the fairness of downstream ML models, as relying solely on one expert may encode subjective biases into the training data. Thus, consensus-based validation is vital for algorithmic fairness in annotated datasets and annotation workflows.

Our evaluation also investigated whether validation outcomes were influenced by the annotator's identity or time spent on the task. Interestingly, neither annotator identity nor annotation time biased the likelihood of acceptance. This contrasts with findings by Rajpurkar et al. (2018), who showed that non-experts crowdsourced annotators achieved diagnostic accuracy comparable to radiologists in chest X-ray interpretation, but required substantially more time to complete the task [7]. One possible explanation is that our structured workflow, with training and feedback loops, may have mitigated variability among annotators. Future studies could investigate how validator feedback and image-specific factors, such as anatomical complexity or noise, affect annotation performance using automated complexity metrics.

Notably, the systematic review by Ørting et al. (2020) also highlights how gold standard annotations are typically treated as ground truth, even though interobserver variability among experts is often overlooked [13]. Out of 21 studies that employed multiple experts for annotation, only three explicitly assessed inter-rater expert agreement. This gap underscores the need for more rigorous validation protocols that account for the inherent subjectivity in medical image interpretation. Integrating measures of expert concordance, and not just comparing crowd annotations to a presumed absolute reference, could provide a more realistic and nuanced benchmark for evaluating annotation quality.

Beyond technical and data contributions, our tool offers educational and socioeconomic benefits. Involving students as annotators eases the demand on radiologists' time while providing structured learning in annotation tasks and anatomical interpretation. As seen in similar initiatives, engaging students in research tasks can support student learning and influence future career interests [9]. The absence of performance disparities across annotators also suggests that training programs can effectively prepare non-experts to contribute to high-quality datasets [4,18], broadening the labor base for equitable AI development.

It is important to acknowledge some limitations of this work. First, we did not evaluate on-premise security requirements for this cloud-based annotation solution. Although cloud systems inherently include security measures, future research could explore this aspect further. Second, we did not analyze potential biases related to the protected attributes (sex and age) of the scanned patients. This represents a promising area for future investigation, as patient profiles may also affect the acceptance of the segmentation. The source code is available for non-profit use upon reasonable request.

In summary, this work shows how collaborative infrastructures can democratize data curation and annotation in medical imaging, a traditionally expertise-intensive and costly domain. With the rise of open-source platforms and collaborative science, initiatives like ours may lower barriers to positively impact medical research, especially in underrepresented regions [16].

References

1. Trame, https://kitware.github.io/trame/
2. Voila: Rendering of live jupyter notebooks with interactive widgets., https://github.com/voila-dashboards/voila
3. Chilamkurthy, S., et al.: Deep learning algorithms for detection of critical findings in head CT scans: a retrospective study. Lancet **392**(10162), 2388–2396 (2018). https://doi.org/10.1016/S0140-6736(18)31645-3, https://linkinghub.elsevier.com/retrieve/pii/S0140673618316453
4. Damgaard, C., Eriksen, T.N., Juodelyte, D., Cheplygina, V., Jiménez-Sánchez, A.: Augmenting chest x-ray datasets with non-expert annotations. arXiv preprint arXiv:2309.02244 (2023)
5. Fedorov, A., et al.: 3D slicer as an image computing platform for the quantitative imaging network. Magn. Reson. Imaging **30**(9), 1323–1341 (2012). https://doi.org/10.1016/j.mri.2012.05.001, https://linkinghub.elsevier.com/retrieve/pii/S0730725X12001816
6. Gulshan, V., et al.: Development and validation of a deep learning algorithm for detection of diabetic retinopathy in retinal fundus photographs. jama **316**(22), 2402–2410 (2016)
7. Heim, E., et al.: Large-scale medical image annotation with crowd-powered algorithms. J. Med. Imaging **5**(3), 034002 (2018)
8. Holzinger, A., et al.: Interactive machine learning: experimental evidence for the human in the algorithmic loop: a case study on ant colony optimization. Appl. Intell. **49**, 2401–2414 (2019)
9. Kraft, M., Sayfie, A., Klein, K., Gruppen, L., Quint, L.: Introducing first-year medical students to radiology: implementation and impact (2018)
10. Lowekamp, B.C., Chen, D.T., Ibáñez, L., Blezek, D.: The Design of SimpleITK. Front. Neuroinf. **7** (2013). https://doi.org/10.3389/fninf.2013.00045

11. McBee, P., Zulqarnain, F., Syed, S., Brown, D.E.: Image-level uncertainty in pseudo-label selection for semi-supervised segmentation. In: Annual International Conference of the IEEE Engineering, pp. 4740–4744 (2022)
12. Nowak, S., Rüger, S.: How reliable are annotations via crowdsourcing: a study about inter-annotator agreement for multi-label image annotation. In: Proceedings of the International Conference on Multimedia Information Retrieval, MIR 2010, pp. 557–566. ACM, New York, NY, USA (2010). https://doi.org/10.1145/1743384.1743478
13. Ørting, S., et al.: A survey of crowdsourcing in medical image analysis. arXiv preprint arXiv:1902.09159 (2019)
14. Pinto, B.G.G., et al.: Clinical validation of a deep learning model forsegmenting and quantifying intracranial andventricular volumes on computed tomography. Sci. Rep. (2024)
15. ScienceDirect: tumor segmentation - an overview — sciencedirect topics, https://www.sciencedirect.com/topics/computer-science/tumor-segmentation, Accessed 20 June 2025
16. Tripathi, S., et al.: Understanding biases and disparities in radiology ai datasets: a review. J. Am. Coll. Radiol. **20**(9), 836–841 (2023)
17. Wang, S., et al.: Annotation-efficient deep learning for automatic medical image segmentation. Nat. Commun. **12**(1), 5915 (2021)
18. Zhang, J., Zheng, Y., Hou, W., Jiao, W.: Leveraging non-expert crowdsourcing to segment the optic cup and disc of multicolor fundus images. Biomed. Opt. Express **13**(7), 3967–3982 (2022)

meval: A Statistical Toolbox for Fine-Grained Model Performance Analysis

Dishantkumar Sutariya and Eike Petersen[✉]

Fraunhofer Institute for Digital Medicine MEVIS, Bremen, Germany
eike.petersen@mevis.fraunhofer.de

Abstract. Analyzing machine learning model performance stratified by patient and recording properties is becoming the accepted norm and often yields crucial insights about important model failure modes. Performing such analyses in a statistically rigorous manner is non-trivial, however. Appropriate performance metrics must be selected that allow for valid comparisons between groups of different sample sizes and base rates; metric uncertainty must be determined and multiple comparisons be corrected for, in order to assess whether any observed differences may be purely due to chance; and in the case of intersectional analyses, mechanisms must be implemented to find the most 'interesting' subgroups within combinatorially many subgroup combinations. We here present a statistical toolbox that addresses these challenges and enables practitioners to easily yet rigorously assess their models for potential subgroup performance disparities. While broadly applicable, the toolbox is specifically designed for medical imaging applications. The analyses provided by the toolbox are illustrated in two case studies, one in skin lesion malignancy classification on the ISIC2020 dataset and one in chest X-ray-based disease classification on the MIMIC-CXR dataset.

Keywords: Model evaluation · Bias assessment · Statistical methods

1 Introduction

It is increasingly well-recognized that developers should assess the performance of their machine learning models not just in the aggregate, but also dis-aggregated – or *stratified* – by attributes characterizing model inputs and targets [7,25,27]. The motivation for such stratified analyses is twofold. Firstly, they enable the identification of *quality-of-service (QoS) biases*, i.e., whether models perform differently well in different patient cohorts. Secondly, and more broadly, such fine-grained analyses can reveal important model failure modes, such as shortcut learning or failures on specific recording devices, field strengths, protocols, or other technical parameters [15,22]. Such shortcomings are often not readily apparent in aggregate test-set performance analyses, and they can also be a cause of QoS biases [26]. In essence, finely stratified performance analyses allow answering the crucial question: *Does this model work for every patient?*

Performing such stratified model performance analyses in a statistically rigorous manner is far from trivial, however. Subgroups may differ strongly in sample size and base rates (incidences), rendering standard performance assessment tools such as precision-recall (PR) curves (and the area below them) as well as the expected calibration error (ECE) and its variants inapplicable [27]. To assess the uncertainty of performance evaluations in subgroups, the computation of reliable uncertainty estimates (confidence intervals) is indispensible, and appropriate statistical tests must be designed while accounting for the fact that potentially *many* subgroups are analyzed and there is, thus, a high risk of false positive findings. Much of the existing literature on performance evaluation methodology focuses on the case of aggregate evaluation [7], providing little guidance on proper methodology for stratified analyses. While there is a large branch of literature on comparative model evaluation [30,31,40], these works mostly focus on the case of *model comparison* which differs crucially: in that setting, the samples of interest (the same test set analyzed using different models) are *paired*, with important implications for proper statistical methodology.

Here, we present a statistical toolbox designed to address these challenges. Our aim is to empower medical imaging practitioners to rigorously assess their models with respect to intersectional subgroup performance disparities.

2 Related Work

Prior work has investigated the limited applicability of commonly-used metrics. The precision-recall (PR) curve, often recommended in the case of strong class imbalance and widely used in the medical domain [40], depends on the base rate $p(y = 1)$ of the sample under test [3,11,40] and is therefore not meaningfully comparable between samples with different base rates [27].[1] This also affects derived metrics such as the area under the precision-recall curve (AUPR), sometimes also called Average Precision (AP), and the geometric mean of precision and recall, the F_1 score. Several alternative base rate-independent metrics have been proposed [3,11]. In terms of calibration measurement, the expected calibration error (ECE) – the most commonly used metric – suffers from a strong sample size bias [4,14,20,33]. Its value for an identically calibrated model thus changes as a function of the test sample size, rendering this metric ill-suited for subgroup comparisons [27,32]. Debiased alternative metrics have been proposed [10,20,33].

Addressing the challenge of identifying potential biases in combinatorially many intersectional subgroups, Kearns et al. [18] and Zhang et al. [45] provide efficient algorithms for specific performance metrics. The extension of these approaches to commonly used metrics such as AUROC is not obvious, however. Most closely related to our work, Cherian et al. [6] recently note that proper statistical testing in the fairness auditing scenario is under-addressed;

---
[1] This issue does not affect the AUROC metric, which is base rate-independent.

they develop a comprehensive and rigorous statistical approach to certifying subpopulation performance disparities. Their approach is highly methodical in nature and deviates from standard model evaluation workflows, limiting accessibility to medical imaging practitioners. Finally, DiCoccio et al. [8] describe a generalized approach to statistical hypothesis testing for arbitrary metrics in the fairness auditing case; we implement this approach in our toolbox.

Toolboxes such as AIF360 [2] and fairlearn [41] may appear to provide similar functionality, but they do not address the specific needs of comprehensive intersectional model evaluation: many standard model performance metrics are not available, and neither statistical testing methodology nor stratified analyses of common performance curves (ROC, PR, calibration) are provided.

3 Methodology

3.1 Metric Choices

We implement standard performance metrics including (balanced) accuracy, AUROC, (balanced) Brier score, sensitivity and specificity. Any metric that is an average over per-recording metric values (such as the average dice score) is implemented via a blanket 'AverageMetric', providing full support in terms of confidence intervals and statistical testing. In addition, we also provide implementations of several non-standard metrics in the toolbox. Most notably, we implement the (partial) area under the precision-recall-gain curve (pAUPRG), originally proposed by Flach et al. [11] and the debiased root mean squared calibration error (DRMSCE) proposed by Petersen et al. [27], which represents an improved version of the debiased estimator proposed by Kumar et al. [20].

The AUPRG metric was originally proposed by Flach et al. [11] to address several noted deficiencies of the AUPR (or AP) metric, including its base rate dependence. Flach et al. define the precision and recall *gains* as

$$precG = \frac{prec - br}{(1 - br)prec} \quad \text{and} \quad recG = \frac{rec - br}{(1 - br)rec} \qquad (1)$$

where $br = P(y = 1)$ denotes the base rate of the test sample. Flach et al. also provided an implementation of their proposed metrics and the area below PRG curves.[2] This implementation has been unmaintained for many years, however, and suffers from several long-known issues, warranting an up-to-date reimplementation. In addition, we discovered a previously undescribed problem with the original method of calculating AUPRG, which we will describe in the following.

The AUPRG is obtained by integrating over the PRG curve from $recG = 0$ to $recG = 1$. $recG = 0$ corresponds to $rec = br$, so this integration requires there to be a well-defined $(recG, precG)$ point at $rec = br$. If there is no decision threshold that happens to yield exactly $rec = br$, this point can be obtained by linear interpolation (which is meaningful in PRG space, unlike in PR space [11])

---

[2] https://github.com/meeliskull/prg.

if and only if there are well-defined points on either side of $rec = br$. 'Well-defined' here refers, in particular, to the value of $prec$, which is only defined if there is at least one positive prediction. The smallest rec value for which $prec$ is well-defined is thus given by the decision threshold corresponding to the highest score value predicted by the model. If the highest predicted score is for a *negative* example, this yields a valid point at $rec = 0$ and $prec = 0$, and we can thus obtain the PRG point at $recG = 0$ by linear interpolation. Alternatively (and more likely), if there is at least one *positive* example in the set of samples obtaining the highest score, we obtain the first well-defined point at

$$rec = \frac{TP}{TP + FN} = \frac{\text{num positives at highest score}}{\text{all positives}}. \quad (2)$$

Especially for small samples sizes and strong class imbalance ($br \ll 0.5$), this point will often be at $rec > br$, rendering AUPRG ill-defined. For this reason, we also provide an implementation of a *partial* AUPRG that is obtained by integrating $precG$ over $[recG_{\min}, 1]$ with $0 \leq recG_{\min} < 1$.

For all metrics derived from an underlying *curve*, such as AUROC, AUPR(G), and DRMSCE (derived from the calibration curve), we also present the corresponding curves split by groups, and confidence intervals obtained by bootstrapping [1]. The operating points selected by a given decision threshold for the different groups are highlighted.

3.2 Intersectional Analyses

Performance disparities should not only be evaluated between groups defined by a single attribute (gender) but also between groups defined by the intersection of multiple attributes (gender × age × technical parameters × ...) [18]. This presents its own challenges: there are combinatorially many subgroups to consider, some of which will be very small (further increasing both the importance and the difficulty of valid metric uncertainty quantification). Taking a pragmatic approach, we simply allow the user to set a minimum group membership threshold for a subgroup to be considered, as well as a maximum interaction level of attributes. For visualization purposes, we select the most 'interesting' subgroups to display based on the sum of a group's ranks in terms of the p-value attached to its performance disparity and the magnitude of that performance disparity, inspired by the 'volcano plots' often used for similar purposes [21].

3.3 Confidence Intervals

All metric results are accompanied by associated confidence intervals (CIs) quantifying the uncertainty about the population-level metric value caused by the fact that the model is evaluated on a small sample drawn from the overall population. For any metric, CIs can be obtained in one of two ways. Firstly, analytical CI approximations can be implemented if they are available for a given metric.

In the current version, we provide analytical CIs for AUROC[3] and ratio-based metrics (Accuracy, Sensitivity, etc.; we use the Wilson score interval implementation of statsmodels [35]). In addition, CIs for any metric can be obtained using a standard percentile bootstrap. In the case of metrics requiring both positive and negative samples (e.g., AUROC) and few samples, we stratify the bootstrap to prevent the excessive occurrence of undefined metric values.

3.4 Statistical Hypothesis Testing

In the case of a single binary attribute of interest (say, gender), it is clear which statistical hypotheses to test for: does model performance differ significantly between these two groups? It is less clear which question to ask (and which hypotheses to test) in the case of subgroups defined by the combination of multiple categorical attributes. Simply testing for pairwise differences between all subgroups results in combinatorially many tests, implying a need to correct for an equally large number of multiple tests and, thus, a high chance of null results even in the presence of non-negligible performance differences. In addition, we are also interested in subgroups of different cardinalities, i.e., groups defined by a different number of attributes. Is it meaningful to compare model performance between, say, 'women' and 'young women'? Finally, choosing specific subgroups to test for significant differences *after* inspecting the results of the model performance assessment would constitute HARKing: 'hypothesizing after the results are known,' a common malpractice closely related to p-hacking [19,38].

In order to circumvent all of these issues, we propose to test for differences with respect to each subgroup's *complementary* group, defined as assuming different values for each group-defining attribute. For instance, if the group under test were defined as 'gender = female and age <25', we would test for differences with respect to the group 'gender not female and age ≥25'. This approach significantly reduces the number of tests to perform compared to the pairwise approach, while still allowing for an exploratory analysis.

To provide a general method for statistical significance testing that is valid for any metric, we implement the approach proposed by DiCiccio et al. [8], in essence a permutation-based test using a studentization of the metric of interest. The studentization requires an estimate of the variance of a given metric's value on a given (permuted) dataset. This variance can be obtained via bootstrapping, but this approach is computationally expensive as it must be repeated for every permutation. We therefore (as also suggested by DiCiccio et al.) use analytical expressions for the variance of a metric wherever they are available. We correct for multiple hypothesis testing using the Holm–Bonferroni correction.

---

[3] Fast DeLong's method [39] as implemented in the 'confidenceintervals' package [12] and a custom implementation of Newcombe's method [24] for small groups (≤ 50 samples) and groups with perfect separation (AUROC = 1.0) since DeLong is known to provide very poor coverage for these cases [9].

3.5 Implementation

The toolbox is implemented in python and designed to be modular, easily extensible, and easy to use. It is publicly available, including all code required to reproduce the two case studies presented below.[4] Visualizations are created using the plotly library, which enables both interactive visualizations and static figure exports. As inputs, the library requires a pandas dataframe with model predictions, ground truth information, and any available metadata. In addition, the user must specify the metrics to analyze. No access to the model is required. Using a single function call, the library creates an interactive HTML report that summarizes model performance across intersectional subgroups. The results are also returned in raw form to enable further custom analyses.

4 Case Studies

We present two case studies to illustrate the kinds of analyses enabled by our toolbox. All plots presented in the following represent direct outputs of simple toolbox function calls, with no further customization applied.

4.1 ISIC Skin Lesion Malignancy Classification

We use the training split of the ISIC2020 dataset [34]. We remove duplicate images based on the list provided on the dataset website and split the data into an 80% training set and a 20% evaluation split, ensuring no lesion leakage based on the 'lesion_id' metadata field.[5] The images are resized to 256 × 256 pixels, center-cropped to 224×224 pixels, and normalized using the parameters provided in the torchvision documentation for the ImageNet-pretrained ResNet50-V2. We finetune the model for 25 epochs using stochastic gradient descent (binary cross-entropy, learning rate 5×10^{-4}, momentum 0.9, batch size 64), random flips and random color jitter (torchvision 0.22.1, all parameters set to 0.25).

Figure 1 shows a (static version of the otherwise dynamic) metric overview generated using our toolbox. Classification accuracy differs significantly between subgroups, but this is apparently primarily a function of the respective subgroups' base rates: no statistically significant AUROC differences are found. We also observe that the model is very poorly calibrated overall.

4.2 MIMIC-CXR Lung Disease Diagnosis

We use the MIMIC-CXR-JPG database [13,16,17], discarding lateral recordings and keeping only frontal (AP/PA) recordings. We discard the 'support device', 'fracture' and 'pleural other' labels, focusing our analyses on the remaining 10 disease labels and the 'No finding' label. Following the approach of Weng et al. [42], we discard multiple recordings for the same patient and keep just one

---

[4] https://github.com/FraunhoferMEVIS/meval.
[5] Patient leakage can still occur [5]; we deemed this to be non-critical for our study.

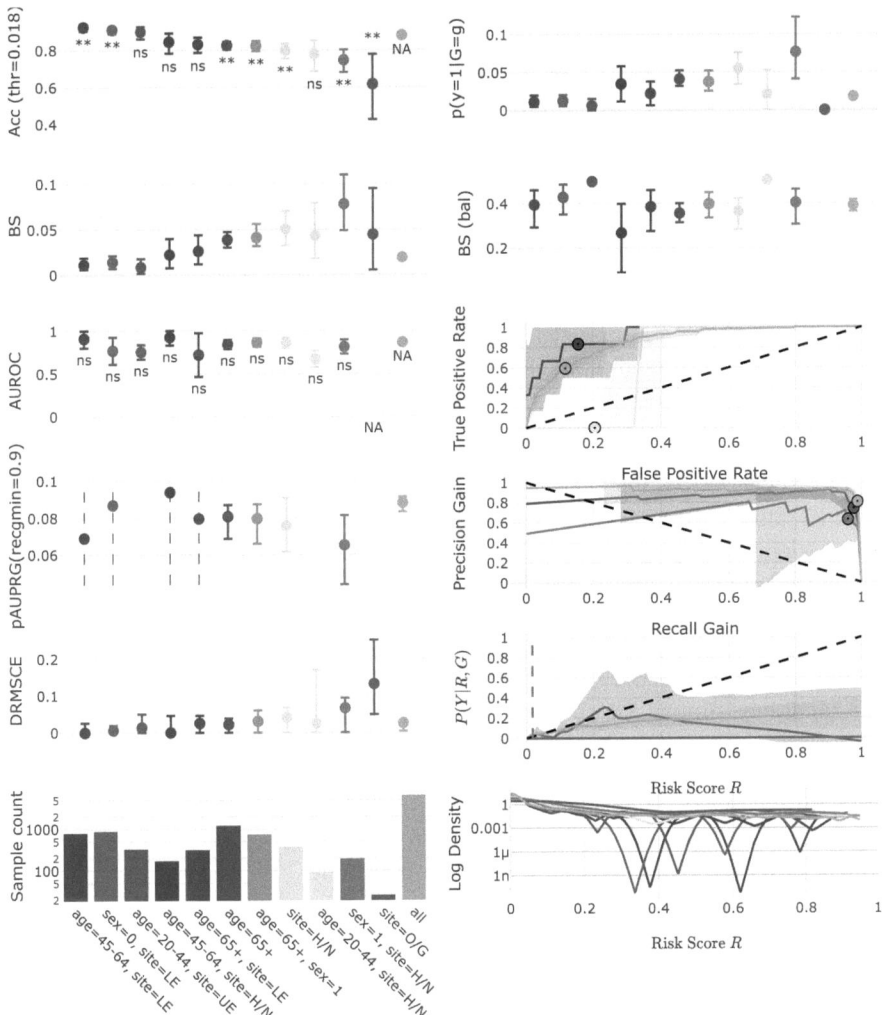

Fig. 1. Exemplary default output of the toolbox for the ISIC2020 case study. Statistical tests performed for accuracy and AUROC. For thresholded metrics, the base rate was chosen as the decision threshold. H/N: head/neck, LE: lower extremity, UE: upper extremity, O/G: oral/genital, TO: torso. ns: not significant ($p > 0.01$), *: $p \leq 0.01$, **: $p \leq 0.001$. Dashed vertical lines indicate that CIs could not be obtained

out of the set with the most disease labels, in order to minimize the risk of label errors [44]. From the resulting dataset of 41,168 recordings, a test set is constructed by randomly sampling 35 positive instances for each of the 11 labels for each of the top-5 race groups, resulting in a total test-set size of 1,757

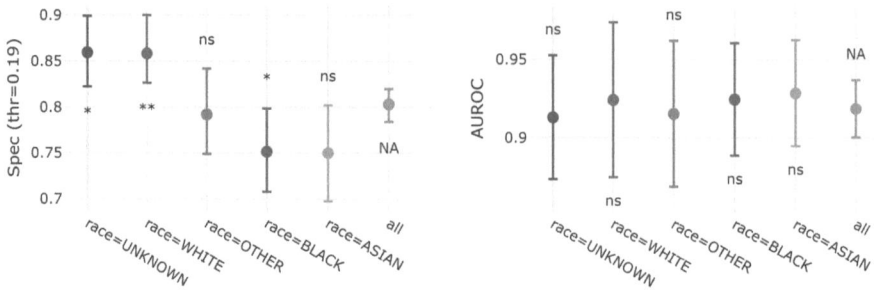

Fig. 2. No-Finding Specificity (left) and AUROC (right), stratified by racial groups. For specificity, the threshold was chosen as in [37] to maximize the geometric mean of sensitivity and specificity. ns: not significant ($p > 0.01$), *: $p \leq 0.01$, **: $p \leq 0.001$

samples.[6] The remaining data are randomly split into a training and validation set of 37,439 (95%) and 1,972 (5%) samples, respectively. We ensure that there is no patient overlap between any of the three sets. We fine-tune a DenseNet121 for multilabel classification, similar to prior work [36,43].

As preprocessing steps, the data are downscaled to 224 × 224 pixels, randomly rotated (max. ±10°), randomly cropped and resized to 200 × 200 pixels and randomly flipped (horizontally and vertically). We optimize binary cross-entropy using AdamW with early stopping (patience 5) based on the validation AUROC. The best (in terms of validation AUROC) intermediate checkpoint is retained, which achieves a macro-averaged validation AUROC of 0.84, and macro-averaged test AUROC of 0.79. Notice that our test set is specifically constructed to be highly diverse and challenging, and a drop in AUROC between the training/validation and test sets is thus to be expected. For our subgroup analyses, following Seyyed-Kalantari et al. [37], we focus on the 'No finding' label and racial groups. Figure 2 shows the per-subgroup 'No-finding' FPR and AUROC.

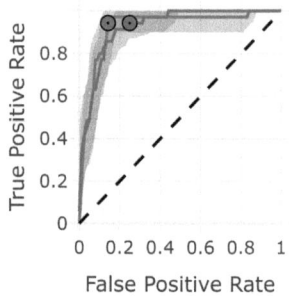

Fig. 3. ROC curves and operating points for white (red) and black (purple) patients (Color figure online)

We reproduce the finding of Seyyed-Kalantari et al. that there is a significant gap in the 'No Finding' specificity between racial groups. Interestingly, however, we observe that there is no significant difference between these groups in terms of AUROC. A comparison of the ROC curves (Fig. 3) indicates that while the overall ROC curves (and the areas under them) are similar, the racial groups are in different ROC operating points (TPR/FPR) for the same decision threshold.

---

[6] For some race-label combinations, less than 35 samples were available. In those instances, we used all available samples in the test set. Notice also that due to the multi-label nature of the data, there may be more than 35 samples from a given race-label combination in the resulting test set.

5 Conclusion and Outlook

Our aim with this work is to provide a statistical toolbox that enables practitioners to conduct rigorous intersectional performance disparity analyses. To this end, we implement several non-standard performance metrics, best-practice statistical methodology, and interactive visualizations for exploring potential disparities. Future work may include the development of a deconfounding approach for disentangling the effects of different causal factors on model performance [23, 28, 29]. We hope that the publication of our toolbox may inspire many researchers to perform case studies in different fields of application, aiding the identification of model blind spots and unfair biases.

Acknowledgments. The authors would like to thank Dr. Max Westphal and M.Sc. David Pfrang for helpful discussions on the statistical methodology. Part of the work that led to this publication has received funding from the European Union's Horizon Europe research and innovation programme under grant agreement No. 101057091.

Disclosure of Interests. The authors declare no competing interests.

References

1. Austin, P.C., Steyerberg, E.W.: Bootstrap confidence intervals for loess-based calibration curves **33**(15), 2699–2700 (2014)
2. Bellamy, R.K.E., Dey, K., Hind, M., Hoffman, S.C., et al.: AI Fairness 360: an extensible toolkit for detecting, understanding, and mitigating unwanted algorithmic bias (2018), https://arxiv.org/abs/1810.01943
3. Boyd, K., Costa, V.S., Davis, J., Page, C.D.: Unachievable region in precision-recall space and its effect on empirical evaluation. In: Proceedings of the 29th International Conference on Machine Learning, pp. 1619–1626 (2012)
4. Bröcker, J.: Estimating reliability and resolution of probability forecasts through decomposition of the empirical score **39**(3–4), 655–667 (2011)
5. Cassidy, B., et al.: Analysis of the ISIC image datasets: usage, benchmarks and recommendations. Med. Image Anal. **75**, 102305 (2022)
6. Cherian, J.J., Candès, E.J.: Statistical inference for fairness auditing. J. Mach. Learn. Res. **25**(149), 1–49 (2024)
7. Collins, G.S., et al.: Evaluation of clinical prediction models (part 1): from development to external validation. BMJ p. e074819 (2024)
8. DiCiccio, C., Vasudevan, S., Basu, K., Kenthapadi, K., Agarwal, D.: Evaluating fairness using permutation tests, pp. 1467–1477, August 2020
9. Feng, D., et al.: A comparison of confidence/credible interval methods for the area under the ROC curve for continuous diagnostic tests with small sample size. Stat. Methods Med. Res. **26**(6), 2603–2621 (2015). https://doi.org/10.1177/0962280215602040
10. Ferro, C.A.T., Fricker, T.E.: A bias-corrected decomposition of the Brier score. Q. J. Royal Meteorol. Soc. **138**(668), 1954–1960 (2012). https://doi.org/10.1002/qj.1924
11. Flach, P., Kull, M.: Precision-recall-gain curves: PR analysis done right. In: Advances in Neural Information Processing Systems, vol. 28 (2015)

12. Gildenblat, J.: A python library for confidence intervals. https://github.com/jacobgil/confidenceinterval (2023)
13. Goldberger, A.L., Amaral, L.A., Glass, L., Hausdorff, J.M., Ivanov, P.C., et al.: PhysioBank, PhysioToolkit, and PhysioNet. Circulation **101**(23), e215–e220 (2000)
14. Gruber, S.G., Buettner, F.: Better uncertainty calibration via proper scores for classification and beyond. In: Advances in Neural Information Processing Systems (2022)
15. Jiménez-Sánchez, A., Juodelyte, D., Chamberlain, B., Cheplygina, V.: Detecting shortcuts in medical images – a case study in chest x-rays. In: International Symposium on Biomedical Imaging (ISBI), IEEE, April 2023
16. Johnson, A., Lungren, M., Peng, Y., Lu, Z., et al.: MIMIC-CXR-JPG - chest radiographs with structured labels. PhysioNet (2024). https://doi.org/10.13026/JSN5-T979
17. Johnson, A., Pollard, T.J., et al.: MIMIC-CXR, a de-identified publicly available database of chest radiographs with free-text reports. Sci. Data **6**(1) (2019)
18. Kearns, M., Neel, S., Roth, A., Wu, Z.S.: Preventing fairness gerrymandering: auditing and learning for subgroup fairness. In: Proceedings of the 35th International Conference on Machine Learning (2018)
19. Kerr, N.L.: HARKing: hypothesizing after the results are known. Pers. Soc. Psychol. Rev. **2**(3), 196–217 (1998). https://doi.org/10.1207/s15327957pspr0203_4
20. Kumar, A., Liang, P.S., Ma, T.: Verified uncertainty calibration. In: Advances in Neural Information Processing Systems, vol. 32. Curran Associates, Inc
21. Li, W., et al.: Using volcano plots and regularized-chi statistics in genetic association studies. Comput. Biol. Chem. **48**, 77–83 (2014)
22. Lotter, W.: Acquisition parameters influence AI recognition of race in chest x-rays and mitigating these factors reduces underdiagnosis bias. Nat. Commun. **15**(1) (2024)
23. Mukherjee, P., et al.: Confounding factors need to be accounted for in assessing bias by machine learning algorithms **28**(6), 1159–1160
24. Newcombe, R.G.: Confidence intervals for an effect size measure based on the Mann–Whitney statistic. Part 2: asymptotic methods and evaluation. Stat. Med. **25**(4), 559–573 (2005). https://doi.org/10.1002/sim.2324
25. Oakden-Rayner, L., Dunnmon, J., Carneiro, G., Re, C.: Hidden stratification causes clinically meaningful failures in machine learning for medical imaging. In: Proceedings of the ACM Conference on Health, Inference, and Learning (2020)
26. Olesen, V., et al.: Slicing through bias: explaining performance gaps in medical image analysis using slice discovery methods. In: MICCAI FAIMI Workshop (2024). https://doi.org/10.1007/978-3-031-72787-0_1
27. Petersen, E., et al.: On (assessing) the fairness of risk score models. In: Proceedings of the 2023 ACM Conference on Fairness, Accountability, and Transparency (2023). https://doi.org/10.1145/3593013.3594045
28. Petersen, E., Holm, S., Ganz, M., Feragen, A.: The path toward equal performance in medical machine learning. Patterns **4**(7) (2023). https://doi.org/10.1016/j.patter.2023.100790
29. Pfohl, S.R., et al.: Understanding challenges to the interpretation of disaggregated evaluations of algorithmic fairness (2025), https://arxiv.org/abs/2506.04193
30. Rainio, O., Teuho, J., Klén, R.: Evaluation metrics and statistical tests for machine learning. Sci. Rep. **14**(1) (2024)
31. Raschka, S.: Model evaluation, model selection, and algorithm selection in machine learning, November 2018. https://doi.org/10.48550/ARXIV.1811.12808

32. Ricci Lara, M.A., Mosquera, C., Ferrante, E., Echeveste, R.: Towards unraveling calibration biases in medical image analysis. In: MICCAI FAIMI Workshop (2023)
33. Roelofs, R., Cain, N., Shlens, J., Mozer, M.C.: Mitigating bias in calibration error estimation. In: Proceedings of The 25th International Conference on Artificial Intelligence and Statistics, vol. 151, pp. 4036–4054 (2022)
34. Rotemberg, V., et al.: A patient-centric dataset of images and metadata for identifying melanomas using clinical context. Sci. Data **8**(1) (2021)
35. Seabold, S., Perktold, J.: statsmodels: Econometric and statistical modeling with python. In: 9th Python in Science Conference (2010)
36. Seyyed-Kalantari, L., Liu, G., McDermott, M., Chen, I.Y., Ghassemi, M.: CheXclusion: fairness gaps in deep chest X-ray classifiers. In: Biocomputing 2021
37. Seyyed-Kalantari, L., Zhang, H., McDermott, M.B.A., Chen, I.Y., Ghassemi, M.: Underdiagnosis bias of artificial intelligence algorithms applied to chest radiographs in under-served patient populations. Nat. Med. **27**(12), 2176–2182 (2021)
38. Stefan, A.M., Schönbrodt, F.D.: Big little lies: a compendium and simulation ofphacking strategies. Royal Soc. Open Sci. **10**(2) (2023). https://doi.org/10.1098/rsos.220346
39. Sun, X., Xu, W.: Fast implementation of DeLong's algorithm for comparing the areas under correlated receiver operating characteristic curves. IEEE Signal Process. Lett. **21**(11), 1389–1393 (2014)
40. Varoquaux, G., Colliot, O.: Evaluating machine learning models and their diagnostic value. In: Machine Learning for Brain Disorders. Springer US (2023). https://doi.org/10.1007/978-1-0716-3195-9_20
41. Weerts, H., et al.: Fairlearn: assessing and improving fairness of AI systems (2023), http://jmlr.org/papers/v24/23-0389.html
42. Weng, N., et al.: Are sex-based physiological differences the cause of gender bias for chest X-ray diagnosis? In: MICCAI 2023 FAIMI Workshop, pp. 142–152 (2023)
43. Yang, Y., Zhang, H., Gichoya, J.W., Katabi, D., Ghassemi, M.: The limits of fair medical imaging AI in real-world generalization. Nat. Med. (2024). https://doi.org/10.1038/s41591-024-03113-4
44. Zhang, H., Dullerud, N., Roth, K., et al.: Improving the fairness of chest X-ray classifiers. In: Conference on Health, Inference, and Learning (CHIL) (2022)
45. Zhang, Z., Neill, D.B.: Identifying significant predictive bias in classifiers. In: NeurIPS FAT/ML Workshop (2017)

Revisiting the Evaluation Bias Introduced by Frame Sampling Strategies in Surgical Video Segmentation Using SAM2

Utku Ozbulak[1,2(✉)], Seyed Amir Mousavi[1,2], Francesca Tozzi[3,4],
Niki Rashidian[4,5], Wouter Willaert[3,4], Wesley De Neve[1,2],
and Joris Vankerschaver[1,6]

[1] Center for Biosystems and Biotech Data Science, Ghent University Global Campus, Incheon, Republic of Korea
utku.ozbulak@ghent.ac.kr
[2] IDLab, ELIS, Ghent University, Ghent, Belgium
[3] Department of GI Surgery, Ghent University Hospital, Ghent, Belgium
[4] Department of Human Structure and Repair, Ghent University, Ghent, Belgium
[5] Department of HPB Surgery and Liver Transplantation, Ghent University Hospital, Ghent, Belgium
[6] Department of Mathematics, Computer Science and Statistics, Ghent University, Ghent, Belgium

Abstract. Real-time video segmentation is a promising opportunity for AI-assisted surgery, offering intraoperative guidance by identifying tools and anatomical structures. Despite growing interest in surgical video segmentation, annotation protocols vary widely across datasets – some provide dense, frame-by-frame labels, while others rely on sparse annotations sampled at low frame rates such as 1 FPS. In this study, we investigate how such inconsistencies in annotation density and frame rate sampling influence the evaluation of zero-shot segmentation models, using SAM2 as a case study for cholecystectomy procedures. Surprisingly, we find that under conventional sparse evaluation settings, lower frame rates can appear to outperform higher ones due to a smoothing effect that conceals temporal inconsistencies. However, when assessed under real-time streaming conditions, higher frame rates yield superior segmentation stability, particularly for dynamic objects like surgical graspers. To understand how these differences align with human perception, we conducted a survey among surgeons, nurses, and machine learning engineers and found that participants consistently preferred high-FPS segmentation overlays, reinforcing the importance of evaluating every frame in real-time applications rather than relying on sparse sampling strategies. Our findings highlight the risk of evaluation bias that is introduced by inconsistent dataset protocols and bring attention to the need for temporally fair benchmarking in surgical video AI.

Keywords: AI-assisted surgery · Bias in surgical video segmentation · Evaluation bias · SAM2

1 Introduction

Throughout history, surgical practice has been shaped by a series of transformative milestones that have fundamentally improved patient outcomes and expanded treatment possibilities [5]. Major advances include general anesthesia [17], antiseptic techniques [12], and imaging modalities such as X-ray, CT, and MRI, which revolutionized surgical planning [8,11]. The advent of minimally invasive techniques in the 20th century substantially reduced trauma, pain, and hospital stays [14] and more recently, the rise of computer-assisted and robotic surgery has increased surgical precision, reduced invasiveness, and enabled more complex procedures to be performed with greater accuracy [2]. Now, the next major frontier in surgery appears to be the integration of artificial intelligence (AI) at every stage of the surgical process [4].

Intraoperatively, AI-assisted robotic systems have the potential to enhance dexterity, accuracy, and consistency, enabling surgeons to operate with unprecedented control [21]. In particular, recent advancements in computer vision hold immense potential for real-time surgical video analysis, offering intraoperative guidance by identifying optimal dissection planes, highlighting high-risk areas, and marking safe zones for dissection [13]. Among state-of-the-art methods, video object segmentation (VOS) has recently emerged as a critical technology for surgical tool and organ segmentation [18]. Unlike traditional segmentation models that process frames independently, VOS models maintain spatial memory and track objects across frames, ensuring better consistency in real-time applications [15]. The SAM2 model, the state-of-the-art zero-shot segmentation framework, is particularly well-suited for surgical video analysis, as it effectively tracks and segments surgical instruments and anatomical structures without requiring task-specific fine-tuning [10,16].

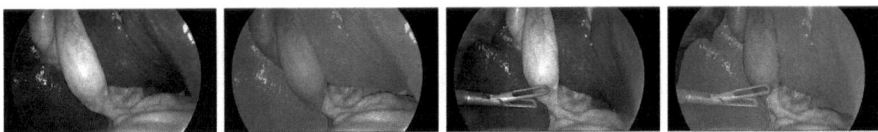

Fig. 1. Two example images from the CholecSeg8k dataset with annotations for the gallbladder (**purple**), liver (light blue), and grasper (green).

Despite the growing interest in surgical video segmentation, annotation protocols differ substantially across datasets. For example, CholecSeg80 [7] provides frame-wise annotations at the native recording rate of 25 FPS, whereas EndoVis 2015 [19], another widely used benchmark, includes labels only at a much sparser 1 FPS setting. CholecSeg8k offers densely annotated semantic masks, but only for 8,080 frames sparsely sampled across multiple videos. In contrast, the CaDIS dataset [6] includes annotations at the native 30 FPS rate. Similarly, CholecT50 [1] provides high-quality per-frame instance masks but still only at 1 FPS, limiting temporal continuity. The variation in annotation strategies and frame-rate settings is so complicated that the authors of [1] dedicated an entire section and two figures just to clarify the annotation protocols and dataset partitions within the Cholec family. These inconsistencies in annotation density and frame sampling introduce potential evaluation biases, as segmentation performance may depend not only on model capability but also on how frequently ground truth labels are provided.

In this work, we evaluate the segmentation performance of the SAM2 model for cholecystectomy on three key targets: the gallbladder, liver, and surgical graspers. We conduct experiments across five frame-per-second (FPS) settings – 1, 10, 15, 20, and 25 FPS – to analyze the impact of frame rate on segmentation accuracy. Our findings reveal an intriguing phenomenon: when evaluated outside of real-time streaming scenarios, higher FPS settings may create the illusion of reduced segmentation performance. In extreme cases, our results show that the **1 FPS setting can outperform 25 FPS** when assessed under conventional evaluation methods. However, when considering real-time scenarios where predictions must remain stable across continuously arriving frames, 25 FPS setting yields superior segmentation performance, ensuring more consistent and temporally coherent segmentation. To complement these findings, we conducted a survey among surgeons, nurses, and machine learning engineers, assessing their perception of segmentation performance at different frame rates. Our findings indicate a strong preference for higher FPS segmentation mask overlays, reinforcing the importance of real-time evaluation in AI-assisted surgery.

2 Methodology

2.1 Model

In this study, we utilize SAM2.1 Hiera Large [16] (henceforth referred to as SAM2), a state-of-the-art zero-shot segmentation model, for real-time surgical video segmentation. SAM2 employs a powerful transformer-based architecture that ensures spatial and temporal consistency, making it particularly well suited for dynamic and complex environments such as surgical video analysis [3,9,20].

Several studies have demonstrated that SAM2 outperforms both traditional image-based segmentation models and other VOS frameworks, achieving state-of-the-art results in surgical video segmentation [22]. A key strength of SAM2 is its built-in tracking mechanism, which enhances temporal consistency across frames, reducing segmentation drift and improving overall coherence. This feature is particularly valuable in the operating room, where maintaining accurate and stable segmentation over time is crucial for real-time decision-making.

2.2 Data

For a comprehensive evaluation across various FPS settings, we use a subset of the CholecSeg8k dataset, a frame-by-frame annotated dataset derived from the widely used Cholec80 surgical video dataset [7,19]. Cholec80 consists of

Table 1. Total number of frames used in each video for segmentation analysis, categorized by target objects (gallbladder, liver, and surgical grasper).

Objects	Video 1	Video 12	Video 17	Video 20	Video 24	Video 35
Any	1,280	640	320	160	960	240
Gallbladder	992	624	320	160	686	240
Liver	1,280	640	320	160	960	240
Grasper	849	577	240	160	400	240

80 cholecystectomy procedure videos recorded at 25 frames-per-second at the University Hospital of Strasbourg (Strasbourg, France) [19]. The CholecSeg8k dataset provides high-quality segmentation annotations for a subset of those videos, making it well-suited for evaluating real-time surgical video segmentation models [7].

In this study, we focus on the segmentation of two key anatomical structures – the gallbladder and liver – as well as a surgical instrument, the grasper. These targets are crucial for surgical scene understanding, as they provide insight into organ visibility and instrument interaction [13]. An example set of images from the dataset is provided in Fig. 1.

Since the employed model is a zero-shot segmentation model, we do not need to partition the dataset into training and validation splits. Instead, we evaluate performance using six videos, some of which are divided into multiple segments. In total, we assess SAM2's performance across five FPS settings – 1, 10, 15, 20, and 25 FPS – for 15 video segments. Here, 25 FPS is the native setting in these videos and corresponds to using all available frames. Table 1 shows the number of frames for each video that contain objects of each class.

2.3 Notation and Metrics

We represent a video as a sequence of n RGB frames, denoted as $\mathbf{X} = [\mathbf{X}_1, \mathbf{X}_2, \ldots, \mathbf{X}_n]$, where each frame $\mathbf{X} \in \mathbb{R}^{3 \times 854 \times 480}$ corresponds to an image with three color channels (RGB) and a resolution of 854 × 480 pixels.

To utilize the SAM2 model for video segmentation, an initial target object must be specified in the first frame, which will then be tracked throughout the sequence. In our approach, we define this target using a binary segmentation mask, represented as $\mathbf{S} \in \{0,1\}^{854 \times 480}$. This mask indicates, for each pixel, whether it belongs (1) or does not belong (0) to the target object.

Given an initial frame $\mathbf{X}_1$ and mask $\mathbf{S}_1$, SAM2 propagates segmentation across subsequent frames, predicting the ith frame as $g(\mathbf{X}_i) = \mathbf{S}_i$. The mask $\mathbf{S}_i$ depends on $\mathbf{S}_1$ and all preceding frames with their inferred segmentations, ensuring temporal consistency in object tracking.

Intersection over Union (IoU). In order to evaluate the correctness of predictions, we employ the IoU metric, which measures the overlap between the

Table 2. (Evaluation: **Sampled frames**) Average IoU scores for gallbladder, liver, and surgical grasper segmentation using the SAM2 model. Results are reported for different frame rates (1, 10, 15, 20, and 25 FPS), where all available frames in each FPS setting are evaluated. The best segmentation performance for each FPS setting per row is highlighted in bold.

VideoSegment	Gallbladder					Liver					Grasper				
	1	10	15	20	25	1	10	15	20	25	1	10	15	20	25
V1 S1	**96.0**	95.7	95.9	95.9	95.9	**90.9**	90.3	90.4	90.2	90.3	**87.8**	86.3	86.4	86.5	86.3
V1 S2	**96.4**	96.1	96.3	96.2	96.1	**97.2**	96.6	97.0	97.1	96.9	**91.2**	86.8	86.7	87.0	86.9
V1 S3	**95.4**	94.9	94.9	94.9	94.9	**96.3**	96.0	96.1	96.2	96.2	**83.9**	82.1	82.2	82.4	82.1
V1 S4	**86.5**	85.5	85.1	84.9	85.2	**96.0**	95.6	95.6	95.6	95.6	**82.3**	80.3	80.2	79.9	80.2
V1 S5	**87.8**	86.7	87.5	86.7	86.8	**93.5**	92.3	92.2	92.2	92.2	78.2	80.4	80.5	80.6	**80.7**
V12 S1	75.4	**82.4**	81.7	81.7	81.6	**92.6**	91.7	91.1	91.2	91.1	78.9	**86.5**	86.4	86.5	86.5
V12 S2	**71.5**	67.0	67.4	68.7	67.7	**94.0**	91.9	91.1	91.4	91.4	54.2	**54.4**	53.7	53.9	53.9
V12 S3	**81.2**	78.2	78.5	77.6	77.9	84.6	**88.5**	88.1	88.3	88.0	55.2	**55.4**	55.3	55.2	55.0
V17 S1	**95.9**	95.5	94.8	94.9	95.0	**95.0**	93.8	92.6	93.1	93.0	**74.5**	69.8	69.7	69.3	69.4
V17 S2	**96.3**	95.1	95.2	95.1	95.2	94.1	92.7	**93.4**	93.1	93.2	**88.5**	86.9	86.7	86.1	86.1
V20 S1	**93.5**	93.1	93.3	93.2	93.0	**96.7**	96.4	96.4	96.4	96.4	**95.1**	94.6	94.6	94.5	94.5
V24 S4	**90.2**	81.4	80.6	78.4	78.3	94.1	**95.4**	95.4	95.4	95.3	**88.0**	87.1	87.2	86.9	86.9
V35 S1	**95.1**	93.7	93.8	93.8	93.7	**98.0**	97.3	97.5	97.6	97.6	**96.2**	94.9	94.1	94.1	94.3
V35 S2	**95.9**	95.6	94.9	94.9	95.0	97.4	**97.8**	97.8	97.8	97.8	**95.1**	93.1	93.3	93.2	93.3
V35 S3	**93.9**	92.1	92.0	92.0	92.0	**98.3**	98.2	98.2	98.2	98.2	**97.0**	96.3	96.2	96.2	96.2

predicted segmentation mask S and the ground truth mask Y. Formally, IoU is defined as $\text{IoU}(S, Y) = \frac{|S \cap Y|}{|S \cup Y|}$.

2.4 Evaluation

In this work, we evaluate the performance of SAM2 on surgical videos sampled in three settings: sampled frames, anchor frames, and real-time streaming frames. **Sampled Frames.** In this setting, we evaluate the IoU over sampled frames within the selected FPS setting. Formally, given a frame sequence X captured at 25 FPS, the evaluation is conducted over the subset $X^{(f)} = [X_1, X_{1+\Delta f}, X_{1+2\Delta f}, \ldots]$, where $\Delta f = 25/f$ determines the interval between selected frames. This method ensures that higher FPS settings use more frames for evaluation, but it introduces potential bias since different FPS settings involve different numbers of frames. For example, in a 10-second video recorded at 25 FPS, there are $10 \times 25 = 250$ total frames, whereas at 1 FPS, there are only $10 \times 1 = 10$ frames. This means that higher FPS settings provide more temporal details, which can influence the evaluation results. To mitigate this bias, we introduce the anchor frames setting.

Anchor Frames. In this setting, we standardize the number of frames used for evaluation across different FPS settings. Specifically, we define the anchor frame set as $X_{\text{Anchor}} = [X_1, X_{26}, X_{51}, \ldots]$, where only the first frame of each second is used, ensuring that all FPS settings are evaluated on the same number of frames.

Table 3. (Evaluation: **Anchor frames**) Average IoU scores for gallbladder, liver, and surgical grasper segmentation using the SAM2 model when evaluated only on anchor frames (the first frame of each second). The best segmentation performance for each FPS setting per row is highlighted in bold.

VideoSegment	Gallbladder					Liver					Grasper				
	1	10	15	20	25	1	10	15	20	25	1	10	15	20	25
V1 S1	96.3	96.4	96.4	**96.5**	**96.5**	90.9	**91.1**	91.0	90.9	90.6	87.8	87.9	**88.1**	88.0	87.9
V1 S2	**97.2**	**97.2**	**97.2**	**97.2**	**97.2**	**97.2**	**97.2**	**97.2**	**97.2**	**97.2**	**91.2**	90.5	90.2	90.5	90.6
V1 S3	95.4	**95.6**	**95.6**	95.5	95.4	96.3	96.4	96.4	**96.5**	**96.5**	**96.2**	**96.2**	**96.2**	**96.2**	**96.2**
V1 S4	94.1	95.7	**95.8**	**95.8**	95.7	96.0	95.9	**96.0**	95.9	95.9	75.4	77.8	77.5	77.7	**78.1**
V1 S5	93.5	93.3	**93.4**	**93.4**	**93.4**	93.5	93.3	**93.4**	**93.4**	**93.4**	82.3	**82.4**	82.3	82.1	82.1
V12 S1	**95.0**	**95.0**	94.9	94.9	**95.0**	92.6	**92.8**	92.2	92.2	92.2	83.5	83.4	**83.5**	83.4	83.2
V12 S2	96.7	**96.7**	96.6	96.6	**96.7**	94.0	**94.1**	94.0	94.0	93.9	**95.0**	**95.0**	94.9	94.9	**95.0**
V12 S3	94.1	94.0	94.0	93.9	**94.1**	84.6	88.8	87.0	**88.9**	86.9	**89.4**	89.3	89.3	**89.4**	89.1
V17 S1	**95.9**	95.8	95.8	95.8	**95.9**	95.0	**95.1**	95.1	95.1	93.8	95.2	**95.3**	**95.3**	95.2	95.2
V17 S2	96.0	96.0	**96.1**	**96.1**	96.0	**94.1**	**94.1**	**94.1**	**94.1**	**94.1**	**95.3**	**95.3**	**95.3**	95.2	95.1
V20 S1	**96.8**	**96.8**	**96.8**	**96.8**	**96.8**	96.7	**96.8**	**96.8**	**96.8**	**96.8**	**95.3**	**95.3**	**95.3**	95.2	95.1
V24 S4	90.2	80.2	79.9	76.1	76.7	94.1	95.7	**95.8**	**95.8**	95.7	95.7	**95.8**	**95.8**	95.7	95.7
V35 S1	98.3	**98.5**	**98.5**	**98.5**	**98.5**	98.0	**98.2**	**98.2**	**98.2**	**98.2**	96.8	96.8	96.8	96.8	**97.0**
V35 S2	98.0	**98.0**	**98.0**	**98.0**	**98.0**	97.4	**98.0**	**98.0**	**98.0**	**98.0**	95.2	**95.3**	**95.3**	95.2	95.1
V35 S3	97.4	**98.0**	**98.0**	**98.0**	**98.0**	98.3	**98.5**	**98.5**	**98.5**	**98.5**	96.8	96.8	96.8	96.8	**97.0**

This approach eliminates potential biases introduced by varying frame counts and provides a fairer comparison across different FPS configurations.

Real-time Streaming Frames. This setting simulates a real-time streaming scenario where predictions from lower FPS settings must persist across intermediate frames. Regardless of the selected FPS setting, evaluation is conducted on all 25 frames per second. Formally, for an FPS f, predictions S_t at sampled frames are propagated forward across intermediate frames until the next sampled frame appears. That is, for an FPS setting of f, a prediction made at X_t remains unchanged for frames $X_{t+1}, \ldots, X_{t+\Delta f - 1}$ until the next prediction update occurs at $X_{t+\Delta f}$. This setup assesses the practical impact of frame rate on segmentation performance in real-time applications, where predictions must remain valid between updates to prevent flickering.

3 Experimental Results

In this section, we present the evaluation results of the SAM2 model across the three evaluation strategies highlighted in Sect. 2.4: (1) sampled frames, (2) anchor frames, and (3) real-time streaming frames. Results for these settings can be found in Tables 2, 3, and 4, respectively.

Sampled Frames. This setting assesses segmentation accuracy by computing the IoU for the sampled frames available in a given FPS setting. Surprisingly, in most cases, the 1 FPS setting outperforms the 25 FPS setting (see Table 2).

Table 4. (Evaluation: **Real-time streaming**) Average IoU scores for gallbladder, liver, and surgical grasper segmentation using the SAM2 model in a real-time streaming scenario, where lower FPS predictions persist across intermediate frames. The best segmentation performance for each FPS setting per row is highlighted in bold.

VideoSegment	Gallbladder					Liver					Grasper				
	1	10	15	20	25	1	10	15	20	25	1	10	15	20	25
V1 S1	79.8	94.1	95.1	95.5	**95.9**	81.8	89.4	89.9	89.9	**90.3**	30.8	35.9	39.7	42.2	**74.4**
V1 S2	88.5	95.5	95.8	96.0	**96.1**	90.9	95.9	96.4	96.6	**96.9**	57.5	83.3	84.6	85.9	**86.9**
V1 S3	84.2	94.0	94.5	94.7	**94.9**	83.3	94.8	95.5	95.9	**96.2**	63.5	84.4	85.5	86.0	**86.3**
V1 S4	67.7	82.7	83.9	84.2	**85.2**	79.9	94.0	94.8	95.2	**95.6**	37.7	73.5	77.0	79.4	**82.1**
V1 S5	61.8	84.1	86.3	86.2	**86.8**	81.3	91.5	91.9	92.0	**92.2**	37.4	75.4	77.8	79.2	**80.2**
V12 S1	62.1	80.0	81.6	81.5	**81.6**	79.1	90.5	90.8	91.0	**91.1**	34.8	73.9	77.8	79.2	**80.7**
V12 S2	57.8	67.0	67.0	**68.8**	67.7	83.7	90.8	91.2	91.4	**91.4**	41.5	52.6	53.1	53.8	**53.9**
V12 S3	59.5	76.9	77.4	77.8	**77.9**	70.4	86.8	87.5	87.8	**88.0**	38.7	80.5	83.5	84.8	**86.5**
V17 S1	66.2	92.0	93.5	94.4	**95.0**	69.3	90.4	91.6	92.4	**93.0**	30.9	64.0	66.4	67.7	**69.4**
V17 S2	78.6	93.0	94.1	94.6	**95.2**	74.8	91.4	92.3	92.8	**93.2**	34.7	75.9	80.1	83.3	**86.1**
V20 S1	83.5	92.1	92.7	93.0	**93.0**	89.4	95.9	96.2	96.3	**96.4**	79.2	92.7	93.7	94.2	**94.5**
V24 S4	45.1	81.7	83.6	85.0	**86.3**	74.2	90.2	91.3	91.9	**92.4**	**86.9**	83.2	80.6	75.9	31.2
V35 S1	56.3	85.6	89.9	91.8	**93.7**	83.6	95.7	96.7	97.1	**97.6**	29.8	77.3	85.6	90.4	**94.3**
V35 S2	56.2	90.3	92.7	93.9	**95.0**	73.6	95.5	96.9	97.5	**97.8**	25.2	78.8	86.5	89.9	**93.3**
V35 S3	69.0	89.4	90.7	91.3	**92.0**	86.6	97.2	97.7	97.9	**98.2**	45.8	87.8	91.7	94.1	**96.2**

This counterintuitive result arises because lower FPS evaluates performance on significantly fewer frames. By selecting only one frame per second, this setting inherently smooths out segmentation inconsistencies, making the results appear more stable and artificially inflating IoU scores. Higher FPS settings, on the other hand, introduce more frequent updates, which expose even minor segmentation variations. These variations, while not necessarily indicative of poor performance, result in a slight reduction in IoU scores. Our observations suggest that this evaluation setting may not be the most reliable method for assessing real-world segmentation stability, as it can mask critical segmentation inconsistencies during operations that become apparent at higher frame rates.

Anchor Frames. To address potential biases introduced by varying frame counts across FPS settings, the anchor frames setting evaluates segmentation only on the first frame of each second. As shown in Table 3, this setting reveals no apparent superior FPS choice, with segmentation performance remaining relatively similar across all frame rates.

Real-time Streaming Frames. This setting simulates real-world deployment by requiring lower FPS predictions to persist across intermediate frames to prevent flickering. This evaluation reflects practical constraints in real-time applications, where model outputs must remain stable between updates. As illustrated in Table 4, segmentation accuracy consistently favors 25 FPS, particularly for the surgical grasper, which is the object that moves by far the most in surgical operations. Unlike traditional evaluation settings, real-time evaluation clearly

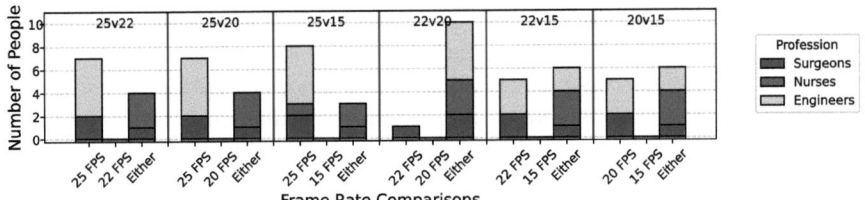

Fig. 2. Illustration of how surgeons, nurses, and engineers perceive differing segmentation frame rates in 25 FPS surgical videos. Each grouped set of bars represents a frame rate comparison between two videos, with responses categorized into three choices: preferring the first video/FPS (left bar), preferring the second video/FPS (middle bar), or no preference/either (right bar).

demonstrates that higher FPS settings are crucial for maintaining segmentation consistency and preventing temporal drift.

3.1 Human Perception of Annotation FPS in Surgical Videos

Following our experiments, we conducted a survey to evaluate the human perception of segmentation frame rates in surgical videos. We asked participants, including three surgeons, three nurses, and four machine learning engineers, to compare videos with segmentation maps generated at different frame rates (e.g., 25v22, 25v20, 25v15) and indicate whether they preferred the higher frame rate, the lower frame rate, or had no preference. Participants were only provided with the segmented video, and did not know the underlying segmentation frame rate. Our goal in this survey is to assess how variations in segmentation FPS impact perceived quality and whether lower FPS affects visual perception.

The survey results provided in Fig. 2 indicate that no participants preferred segmentation maps generated at lower frame rates in any comparison. Instead, when differences were subtle, respondents tended to select the "either is okay" option. As the gap in segmentation FPS increased, more participants noticed a difference, with a preference for higher FPS. Most participants described segmentation mask overlays with lower FPS settings as being "choppy" and out-of-sync.

4 Conclusions

In this work, we revisited the impact of frame rate sampling on zero-shot surgical video segmentation through a fairness lens, focusing on how evaluation strategies can introduce unintended biases in model performance assessment. Using the SAM2 model, we evaluated segmentation accuracy across multiple frame sampling rates and found a counterintuitive result: under conventional evaluation settings, lower frame rates often appear to outperform higher ones. This apparent advantage is largely due to a smoothing effect – fewer segmentation updates result in fewer opportunities for visible errors. However, when models are evaluated under real-time streaming conditions, where predictions must remain temporally consistent across all frames, higher frame rates yield better performance by maintaining segmentation stability and coherence.

To understand how these effects align with human perception, we conducted a survey involving surgeons, nurses, and machine learning engineers. The results highlighted a clear bias in perceptual preference: as the time gap between segmentation mask overlays increased, participants consistently favored higher FPS outputs. This suggests that evaluation protocols based solely on sparse-frame metrics may obscure issues critical to clinical usability and reinforces the importance of temporally aware evaluation in AI-assisted surgery. As such, we call for fairer evaluation practices in surgical video segmentation that move beyond sparse-frame sampling and better reflect both clinical expectations and temporal stability.

References

1. Alabi, O., et al.: Cholecinstanceseg: a tool instance segmentation dataset for laparoscopic surgery. Sci. Data **12**(1), 1–12 (2025)
2. Davies, B.: A review of robotics in surgery. Proc. Inst. Mech. Eng. **214**(1), 129–140 (2000)
3. Dosovitskiy, A., et al.: An image is worth 16x16 words: transformers for image recognition at scale. arXiv preprint arXiv:2010.11929 (2020)
4. Esteva, A., Chou, K., Yeung, S., Naik, N., Madani, A., Mottaghi, A.: Deep learning-enabled medical computer vision. npj Digit. Med. **4**(5) (2021)
5. Gawande, A.: Better: a surgeon's notes on performance. Picador (2012)
6. Grammatikopoulou, M., et al.: Cadis: cataract dataset for surgical rgb-image segmentation. Med. Image Anal. **71**, 102053 (2021)
7. Hong, W.Y., Kao, C.L., Kuo, Y.H., Wang, J.R., Chang, W.L., Shih, C.S.: Cholecseg8k: a semantic segmentation dataset for laparoscopic cholecystectomy based on cholec80. arXiv preprint arXiv:2012.12453 (2020)
8. Hounsfield, G.N.: Computerized transverse axial scanning (tomography): Part 1. description of system. Br. J. Radiol. **46**, 1016–1022 (1973)
9. Kang, S., Vankerschaver, J., Ozbulak, U.: Identifying critical tokens for accurate predictions in transformer-based medical imaging models. In: International Workshop on Machine Learning in Medical Imaging, pp. 169–179. Springer (2024)
10. Kirillov, A., Mintun, E., Ravi, N., et al.: Segment anything. arXiv preprint arXiv:2304.02643 (2023)
11. Lauterbur, P.C.: Image formation by induced local interactions: Examples employing nuclear magnetic resonance. Nature **242**, 190–191 (1973)
12. Lister, J.: On the antiseptic principle in the practice of surgery. Lancet **90**(2299), 353–356 (1867)
13. Maier-Hein, L., Engelhardt, S., Syben, A.M.R., et al.: Computer-assisted medical image analysis for intervention planning: current and future challenges. Med. Image Anal. **33**, 66–75 (2017)
14. Mishra, R.: Minimally Invasive Surgery. Jaypee Brothers Medical Publishers (2008)
15. Oh, S.W., Lee, J.Y., Xu, N., Kim, S.J.: Video object segmentation using space-time memory networks. IEEE Conference on Computer Vision and Pattern Recognition (CVPR), pp. 9226–9235 (2019)
16. Ravi, N., et al.: Sam 2: segment anything in images and videos. arXiv preprint arXiv:2408.00714 (2024)

17. Snow, J.: On chloroform and other anaesthetics: their action and administration (1858)
18. Taghavi, N., Fereshtehnejad, S.M., Navab, N.: Advancements in deep learning-based video object segmentation for medical applications. IEEE Trans. Med. Imaging **41**(12), 3152–3166 (2022)
19. Twinanda, A., Shehata, S., Mutter, D., Marescaux, J., Mathelin, M., Padoy, N.: Endonet: a deep architecture for recognition tasks on laparoscopic videos. IEEE Trans. Med. Imaging **36**(1), 86–97 (2017)
20. Vaswani, A., et al.: Attention is all you need. In: Advances in Neural Information Processing Systems, vol. 30 (2017)
21. Yang, G.Z., Tsang, J.J.C., Martin, J.: From autonomous systems to autonomous robots: the next medical revolution? Proc. IEEE **105**(10), 1954–1966 (2017)
22. Yu, J., et al.: Sam 2 in robotic surgery: An empirical evaluation for robustness and generalization in surgical video segmentation. arXiv preprint arXiv:2408.04593 (2024)

Disentanglement and Assessment of Shortcuts in Ophthalmological Retinal Imaging Exams

Leonor Fernandes[1](✉), Tiago Gonçalves[1], João Matos[2], Luis Nakayama[3], and Jaime S. Cardoso[1]

[1] INESC TEC, Faculdade de Engenharia, Universidade do Porto, Rua Dr. Roberto Frias, 4200-465 Porto, Portugal
{leonor.j.fernandes,jaime.cardoso}@inesctec.pt
[2] University of Oxford, Oxford, UK
[3] Federal University of São Paulo, São Paulo, Brazil

Abstract. Diabetic retinopathy (DR) is a leading cause of vision loss in working-age adults. While screening reduces the risk of blindness, traditional imaging is often costly and inaccessible. Artificial intelligence (AI) algorithms present a scalable diagnostic solution, but concerns regarding fairness and generalization persist. This work evaluates the fairness and performance of image-trained models in DR prediction, as well as the impact of disentanglement as a bias mitigation technique, using the diverse mBRSET fundus dataset. Three models, ConvNeXt V2, DINOv2, and Swin V2, were trained on macula images to predict DR and sensitive attributes (SAs) (e.g., age and gender/sex). Fairness was assessed between subgroups of SAs, and disentanglement was applied to reduce bias. All models achieved high DR prediction performance in diagnosing (up to 94% AUROC) and could reasonably predict age and gender/sex (91% and 77% AUROC, respectively). Fairness assessment suggests disparities, such as a 10% AUROC gap between age groups in DINOv2. Disentangling SAs from DR prediction had varying results, depending on the model selected. Disentanglement improved DINOv2 performance (2% AUROC gain), but led to performance drops in ConvNeXt V2 and Swin V2 (7% and 3%, respectively). These findings highlight the complexity of disentangling fine-grained features in fundus imaging and emphasize the importance of fairness in medical imaging AI to ensure equitable and reliable healthcare solutions.

Keywords: Deep learning · Diabetic retinopathy · Disentanglement · Fairness

1 Introduction

Diabetic retinopathy (DR) is one of the most common complications of diabetes and a leading cause of visual impairment and blindness worldwide. The disease develops when chronically high blood glucose levels damage the small blood vessels that supply the retina, resulting in swelling, leakage or bleeding. In response, the eye may attempt to grow new, but often abnormal and fragile blood vessels that can further impair vision or lead to retinal detachment. In its early stages, the condition is typically asymptomatic. As DR advances, individuals may experience symptoms such as blurry vision and floating spots in their vision. If left untreated, DR can lead to irreversible blindness. Regular eye examinations are essential for people with diabetes, as early detection and timely treatment can slow or prevent the progression of DR and help preserve vision [11].

In this work, we tackled the role of disentanglement in mitigating fairness issues in DR prediction in retinal fundus images, making the following contributions:

1. Analysis of mBRSET, a handheld fundus camera dataset, for group fairness across sensitive attribute (SA) subgroups, using the area under the receiver operating characteristic curve (AUROC), decision curve analysis, and risk distribution plots;
2. Development of a disentangled autoencoder architecture to separate medically relevant features from SA-related information;
3. Test of the architecture as a fairness mitigation technique, by comparing its performance and fairness outcomes against baseline models.

The code related to the implementation of this paper is publicly available in a GitHub repository[1].

2 Background and Related Work

DR datasets are often imbalanced, which can exacerbate shortcut learning, as models may learn dataset-specific associations that do not generalize well to new populations or clinical settings [14]. In medical imaging, SAs are often encoded in the data, allowing models to rely on these features as predictive shortcuts, which can lead to both unfairness and decreased domain generalization [21]. To address these challenges, various bias mitigation techniques have been proposed, including disentanglement learning, adversarial debiasing, and integration of fairness-related penalties [20]. Adding fairness constraints or regularization terms to the loss function is a common approach. However, applying this approach to deep neural networks (DNNs) is challenging, due to their overparameterized nature, which may limit fairness generalization [1].

Disentanglement learning separates the underlying independent factors of variation within data, such as medical features and SAs. This approach can improve predictive performance [16], improve fairness [4], and mitigate shortcut learning [17]. In ophthalmology, disentanglement has been applied to tasks like anonymization and de-identification of retinal images [10,22], as well as domain generalization across imaging devices and clinical sites [2,12,19]. While its potential to improve fairness has been explored in other medical imaging domains [4,17], it remains underexplored in ophthalmology.

---

[1] https://github.com/leofer99/disentanglement_retinal_images.

3 Methodologies

3.1 Dataset and Preprocessing

We used the mBRSET dataset, introduced in [13], the first publicly available dataset of retinal images captured using handheld retinal cameras. It contains 5,164 images from 1,291 patients of diverse ethnic backgrounds in Brasil and includes demographic information such as sex, age, and insurance. We refer the reader to [13] for an in-depth analysis of the remaining attributes.

For the binary classification task, DR severity was binarized using the ICDR scale: **Normal** (levels 0–1) and **Referable** (levels 2–4). Only high-quality, macula-centered images were used to ensure anatomical consistency. All images were resized to 224 × 224 pixels and normalized using ImageNet statistics (i.e., mean and standard deviation) to align with pretrained model expectations [3]. The dataset was split into training (70%), validation (10%), and testing (20%) subsets, stratified by DR distribution and patient identity to prevent data leakage. The DR prevalence in each subset was 18%, 30%, and 17%, respectively. Data augmentation (i.e., random rotation, flips, color jitter, Gaussian blur) was applied during training to improve generalization.

3.2 Models and Training

We trained three image-based deep learning architectures, including a convolutional neural network, ConvNeXt V2 [18], and vision transformer-based models, DINOv2 [15] and Swin V2 [9]. To address class imbalance, we employed focal loss, with class weights computed from the training distribution [6]. Models were trained using the Adam optimizer [5] with early stopping based on F1-score. We added fully connected (FC) layers to project the model embeddings into the output classes. Key hyperparameters included a hidden dimension of 128, a learning rate of 1×10^{-5}, and a batch size of 4 for ConvNeXt V2 and DINOv2, and 2 for Swin V2.

3.3 Fairness Assessment

To evaluate fairness, we used a group fairness approach, focusing on the area under the receiver operating characteristic (AUROC), risk distribution plots, and decision curve analysis (DCA), stratified by groups of patients defined by SAs. The analysis considered five sensitive attributes: age (≤ 50 vs. > 50), sex (female vs. male), education (literate vs. illiterate), insurance (none vs. insured), and obesity (non-obese vs. obese). SAs were selected based on their ability to capture key demographic, socioeconomic, and clinical factors that could contribute to health inequities.

3.4 Disentanglement Architecture

To decrease model bias, we developed a disentanglement architecture that could separate the medical information from the information of a specified SA. This work is based on the architecture proposed in [10], which used a generative model for disentangling identity and medical characteristics in images. The adapted architecture is represented in Fig. 1.

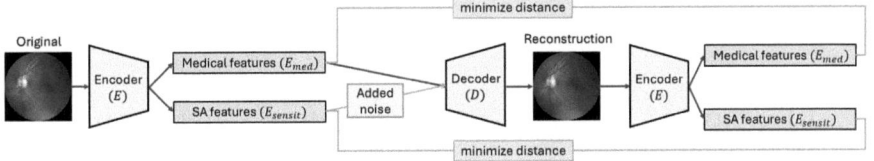

Fig. 1. Overview of the disentanglement network

The four main components of the network are:

- **Encoder** (E): Transforms the input image into two independent vectors, (E_{med}), that encodes the medical features, and (E_{sensit}), that encodes the SA features.
- **Diabetic Retinopathy Classifier** (C_{med}): Guides E_{med} to capture medically relevant information by classifying DR severity.
- **Sensitive Attribute Classifier** (C_{sensit}): The SA classification task is used to approximate the E_{sensit} vector to the geometric space of the SA features.
- **Decoder** (D): Trained to reconstruct the original image from the vectors E_{med} and E_{sensit}, promoting realistic reconstruction and enforcing the utility of both vectors.

The disentangled architecture was trained on the mBRSET dataset with a latent dimension of 256, batch size of 32, a learning rate of 5×10^{-5}, and a weight decay of 1×10^{-6}. The model is trained using a total loss function composed of three weighted components, each with a specific objective. The loss terms are:

- **Disentanglement Loss** ($\mathscr{L}_{\text{disent}}$): Represented in Eq. (1), promotes independence between latent vectors E_{med} and E_{sensit} by penalizing scenarios where modifying a latent vector affects the other. Gaussian noise is added to only one of the latent vectors to simulate perturbations, and an altered image (I_i) is generated.

$$\mathscr{L}_{\text{disent}} = \mathbb{E}_{I \sim p_d(I)} \left[\sum_{i \in \{\text{med},\text{sensit}\}} \left((E_i(I_{\text{ori}}) - E_i(I_i))^2 \right. \right. \\ \left. \left. + \sum_{\substack{j \in \{\text{med},\text{sensit}\} \\ j \neq i}} (E_j(I_{\text{ori}}) - E_j(I_i))^2 \right) \right] \quad (1)$$

- **Classification Loss** ($\mathscr{L}_{\text{classifier}}$): Ensures that E_{med} and E_{sensit} encode the intended features by evaluating their effectiveness on DR and SA classification tasks, respectively. $\mathscr{L}_{\text{classifier}}$ is represented in Eq. (2).

$$\mathscr{L}_{\text{classifier}} = -\lambda_{\text{med}} \sum_c y_{\text{med}}(c) \log(p_{\text{med}}(c)) \\ - \lambda_{\text{sensit}} \sum_k y_{sensit}(k) \log(p_{sensit}(k)) \quad (2)$$

- **Realism Loss** ($\mathscr{L}_{\text{realism}}$): Promotes high-quality image reconstruction by optimizing pixel-level similarity. This loss combines the Structural Similarity Index Measure (SSIM) and Peak Signal-to-Noise Ratio (PSNR), normalized with a threshold parameter $\alpha = 48$ for PSNR. $\mathscr{L}_{\text{realism}}$ is represented in Eq. (3).

$$\mathscr{L}_{\text{realism}} = \mathbb{E}_{I \sim p_d(I)} \left[(1 - \text{SSIM}(I_{\text{ori}}, D(E(I_{\text{ori}})))) + \left(1 - \frac{1}{\alpha} \text{PSNR}(I_{\text{ori}}, D(E(I_{\text{ori}})))\right) \right] \quad (3)$$

- **Total Loss** ($\mathscr{L}_{\text{total}}$): Represented in Eq. (4), combines the three loss terms into a total training objective. The relative contribution of the realism and disentanglement losses is controlled by the weighting factors λ_r and λ_d, which were set to 1 and 5, respectively.

$$\mathscr{L}_{\text{total}} = \mathscr{L}_{\text{classifier}} + \lambda_r \mathscr{L}_{\text{realism}} + \lambda_d \mathscr{L}_{\text{disent}} \quad (4)$$

4 Results and Discussion

4.1 Baseline Results

Classification of Image Models. Among the models, ConvNeXt V2 exhibited the highest overall performance, with an AUROC of 94%, and DINOv2 consistently underperformed, achieving lower AUROC scores and net benefit, particularly in identifying referable DR cases. Models consistently performed better in identifying non-DR than referable DR. This is likely attributable to class imbalance, as referable DR cases only correspond to 17% of mBRSET.

Sensitive Attribute Group Prediction. For the mBRSET dataset, the predicted SAs were age, sex, educational level, insurance, and obesity (see Table 1). Age was predicted well by all models, while sex prediction was moderate, with the exception of DINOv2, which performed poorly. All models struggled to predict educational level, insurance, and obesity.

Table 1. Performance of SA prediction using different models for the test set, reported as AUROC (%) and balanced accuracy (BA) (%).

Model	Age		Sex		Educational Level		Insurance		Obesity	
	AUROC	BA	AUROC	BA	AUROC	BA	AUROC	BA	AUROC	BA
ConvNeXt V2	90.50	69.09	76.60	68.29	56.96	53.57	46.65	50.00	63.14	50.00
DINOv2	90.73	77.32	53.35	50.00	57.01	50.00	48.79	50.00	40.29	50.00
Swin V2	87.85	78.64	76.79	69.89	63.02	52.64	41.74	50.00	58.50	50.00

Fairness Analysis of DR Prediction with mBRSET. Table 2 shows AUROC disparities across models and SA groups. ConvNeXt V2 had the most balanced performance, with AUROC differences typically within 1–3%, indicating strong generalizability and fairness. Swin V2 also performed consistently, with most AUROC disparities remaining under 4%. It exhibited slightly higher discrepancies, particularly between different educational levels and obesity groups. DINOv2 had the largest discrepancies, including age (10%), educational level (7%), and insurance (12%), raising fairness concerns. DeLong's test revealed statistically significant differences only among the obesity subgroups.

Table 2. Performance of DR prediction across SA groups using different models for the test set, reported as AUROC (%) with 95% confidence intervals. Note: A stands for Attributes, S stands for Subgroups, G0 and G1 stand for Group 0 and Group 1, and N is the number of observations in the subgroup. Statistically significant differences between subgroups, as determined by the DeLong's test, are marked with an asterisk (*).

A	S	N	ConvNeXt V2	DINOv2	Swin V2
Full	-	498	94.33 (91.14, 97.53)	88.31 (83.69, 92.93)	90.96 (86.84, 95.07)
Age	G0 (≤ 50)	81	97.06 (93.48, 100.00)	80.09 (63.94, 96.24)	88.69 (75.46, 100.00)
	G1 (> 50)	417	93.79 (90.05, 97.52)	89.81 (85.24, 94.39)	91.41 (87.11, 95.70)
Sex	G0 (female)	329	95.22 (92.16, 98.28)	88.95 (82.76, 95.14)	91.10 (85.51, 96.70)
	G1 (male)	169	93.43 (87.54, 99.33)	88.21 (81.31, 95.10)	90.83 (84.69, 96.96)
Educational	G0 (literate)	222	95.22 (91.22, 99.23)	92.43 (86.10, 98.76)	92.74 (87.19, 98.28)
Level	G1 (illiterate)	268	93.71 (89.12, 98.30)	85.52 (79.06, 91.99)	89.73 (83.99, 95.46)
Insurance	G0 (none)	455	94.26 (90.93, 97.59)	88.86 (84.15, 93.57)	91.27 (87.02, 95.51)
	G1 (insured)	39	97.14 (92.08, 100.00)	77.14 (53.09, 100.00)	85.71 (71.68, 99.74)
Obesity	G0 (non-obese)	450	93.87 (90.35, 97.40)*	87.93 (83.12, 92.74)*	91.10 (87.03, 95.17)*
	G1 (obese)	40	100.00 (100.00, 100.00)*	100.00 (100.00, 100.00)*	100.00 (100.00, 100.00)*

4.2 Disentanglement Architecture

Three representative cases from the mBRSET dataset were selected for analysis based on the highest observed performance disparities and the predictability of SAs. Results are presented in Fig. 2.

ConvNeXt V2 - Disentangling Medical from Age Information. Disentanglement reduced AUROC from 94% to 87%. While age disparity decreased from 4% to 1%, nearly all remaining attributes observed increased disparities, most notably insurance (3% to 21%). DCA showed no clinical utility gain between age groups. Risk distribution plots revealed lower false positives but higher false negatives, explaining the performance drop. It is possible that the model relied on age-related information for DR prediction, as evidenced by the performance drop when this information was removed and its strong age prediction performance.

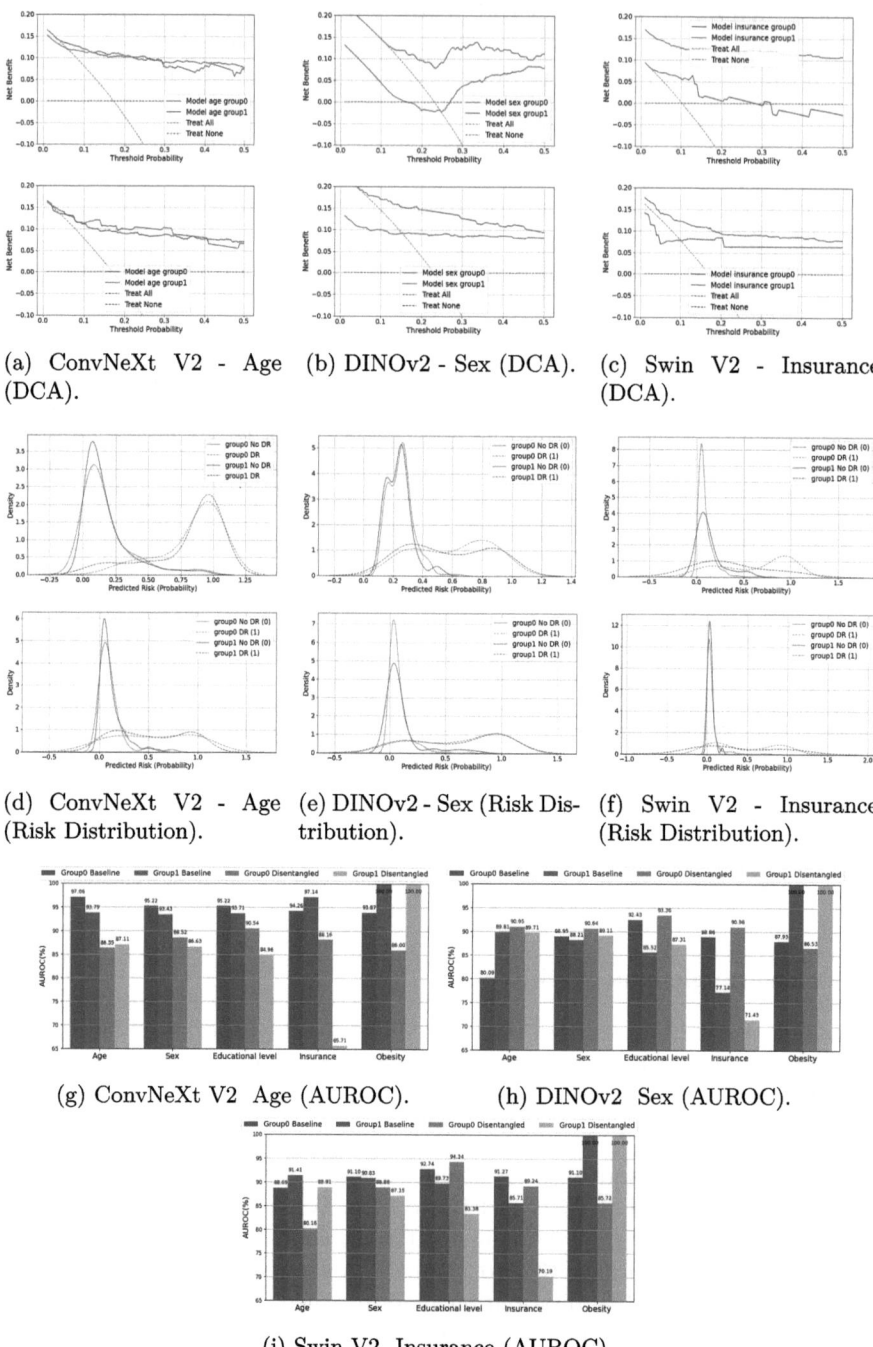

Fig. 2. Performance comparison of baseline and disentangled models for ConvNeXt V2 (Age), DINOv2 (Sex), and Swin V2 (Insurance) on the mBRSET dataset. Rows show: decision curve analysis (top), risk distribution plots (middle), and AUROC disparities across SA groups (bottom).

DINOv2 - Disentangling Medical from Sex Information. Disentanglement improved AUROC from 88% to 90%. Fairness improved for age and educational level (disparities reduced by 8% and 1%), while sex and obesity had minimal increases (both by 1%), and insurance increased by 6%. DCA and risk distribution plots revealed improved clinical utility and reduced false positive and false negative rates. The results also suggest that lowering the decision threshold slightly could benefit the model. Despite weak sex prediction, disentangling sex improved both fairness and generalization.

Swin V2 - Disentangling Medical from Insurance Information. AUROC declined slightly from 91% to 88% with disentanglement. Fairness declined across most SAs, including age, insurance, and obesity (5%, 13%, and 6% respectively). DCA showed a decrease in clinical utility disparity between insurance groups in the disentangled model, while risk distribution plots revealed fewer false positives for insured patients but more false negatives for uninsured patients. Disentanglement offered limited fairness improvement and harmed performance. It is unlikely that the model relied on insurance-encoded information for DR prediction, given its weak performance on the insurance prediction task. In these cases, the model appeared to disentangle irrelevant features, focusing on DR-relevant features, even if this did not align perfectly with the targeted SA.

Summary. The CNN model outperformed the transformer models, consistent with known limitations of transformer-based models on small datasets [7]. It is possible that DINOv2 is overfitting due to the smaller dataset, and disentanglement may help mitigate this issue. The fine-grained nature of retinal features makes disentanglement challenging, as it requires removing SA-related information while preserving fine details, which are necessary for accurate clinical interpretation. This trade-off between removing bias and maintaining diagnostic detail complicates the learning process and can hinder model performance. Moreover, this performance drop may impact subgroups differently, especially for imbalanced ones such as those defined by insurance and obesity, leading to increased performance disparities. Furthermore, DR prediction may inadvertently rely on SA-related information, leading to decreased performance when this information is removed. The relationship between DR prediction and SAs is complex. Improving fairness for one SA does not necessarily imply fairness improvements for other SAs. In fact, optimizing for a specific attribute may increase disparity in others, particularly when attributes are correlated. Additionally, in cases where confounding factors are present, latent dependencies between SAs and DR-relevant features may persist even after standard disentanglement is applied [8], limiting effectiveness.

5 Conclusion and Future Work

Fairness is crucial in AI healthcare to prevent unequal treatment outcomes and should be assessed before clinical implementation. This work investigated fairness in DR detection and explored disentanglement as a mitigation strategy. While disentanglement remains a promising direction, its effectiveness in fundus imaging is limited by domain-specific challenges. Bias mitigation in this context is non-trivial, and more robust techniques may be necessary to improve fairness without compromising predictive performance.

Acknowledgments. This work has received funding from the European Union's Horizon Europe research and innovation programme under the Grant Agreement 101057389-CINDERELLA project.

Disclosure of Interests. The authors have no competing interests to declare that are relevant to the content of this article.

References

1. Cherepanova, V., Nanda, V., Goldblum, M., Dickerson, J.P., Goldstein, T.: Technical challenges for training fair neural networks (2021), https://arxiv.org/abs/2102.06764
2. Chokuwa, S., Khan, M.H.: Generalizing across domains in diabetic retinopathy via variational autoencoders. In: Celebi, M.E., et al. (eds.) Medical Image Computing and Computer Assisted Intervention - MICCAI 2023 Workshops, pp. 265–274. Springer Nature Switzerland, Cham (2023)
3. Deng, J., Dong, W., Socher, R., Li, L.J., Li, K., Fei-Fei, L.: ImageNet: a large-scale hierarchical image database. In: 2009 IEEE Conference on Computer Vision and Pattern Recognition, pp. 248–255. IEEE, Miami, FL, June 2009. https://doi.org/10.1109/CVPR.2009.5206848, https://ieeexplore.ieee.org/document/5206848/
4. Du, S., Hers, B., Bayasi, N., Hamarneh, G., Garbi, R.: FairDisCo: fairer AI in dermatology via disentanglement contrastive learning. In: Karlinsky, L., Michaeli, T., Nishino, K. (eds.) Computer Vision - ECCV 2022 Workshops, pp. 185–202. Springer Nature Switzerland, Cham (2023)
5. Kingma, D.P., Ba, J.: Adam: a method for stochastic optimization (2017), https://arxiv.org/abs/1412.6980
6. Lin, T.Y., Goyal, P., Girshick, R., He, K., Dollár, P.: Focal loss for dense object detection. IEEE Trans. Pattern Anal. Mach. Intell. **42**(2), 318–327 (2020). https://doi.org/10.1109/TPAMI.2018.2858826
7. Linde, G., Rodrigues De Souza Jr, W., Chalakkal, R., Danesh-Meyer, H.V., O'Keeffe, B., Chiong Hong, S.: A comparative evaluation of deep learning approaches for ophthalmology. Sci. Rep. **14**(1), 21829 (2024). https://doi.org/10.1038/s41598-024-72752-x
8. Liu, X., Li, B., Vernooij, M.W., Wolvius, E.B., Roshchupkin, G.V., Bron, E.E.: AI-based association analysis for medical imaging using latent-space geometric confounder correction. Med. Image Anal. **102**, 103529 (2025). https://doi.org/10.1016/j.media.2025.103529. May
9. Liu, Z., et al.: Swin transformer: hierarchical vision transformer using shifted windows. In: 2021 IEEE/CVF International Conference on Computer Vision (ICCV), pp. 9992–10002 (2021). https://doi.org/10.1109/ICCV48922.2021.00986
10. Montenegro, H., Cardoso, J.S.: Anonymizing medical case-based explanations through disentanglement. Med. Image Anal. **95**, 103209 (2024). https://doi.org/10.1016/j.media.2024.103209. https://linkinghub.elsevier.com/retrieve/pii/S1361841524001348

11. Mounirou, B.A.M., Adam, N.D., Yakoura, A.K.H., Aminou, M.S.M., Liu, Y.T., Tan, L.Y.: Diabetic retinopathy: an overview of treatments. Indian J. Endocrinol. Metab. **26**(2), 111–118 (2022). https://doi.org/10.4103/ijem.ijem_480_21, https://journals.lww.com/10.4103/ijem.ijem_480_21
12. Müller, S., Koch, L.M., Lensch, H.P., Berens, P.: Disentangling representations of retinal images with generative models. Med. Image Anal. **105**, 103628 (2025). https://doi.org/10.1016/j.media.2025.103628
13. Nakayama, L.F., et al.: mBRSET, a mobile Brazilian retinal dataset. https://doi.org/10.13026/QXPD-1Y65, https://physionet.org/content/mbrset/1.0/
14. Ong Ly, C., et al.: Shortcut learning in medical ai hinders generalization: method for estimating ai model generalization without external data. npj Digital Medicine **7**(1), 1–10 (2024). https://doi.org/10.1038/s41746-024-01118-4, https://www.nature.com/articles/s41746-024-01118-4
15. Oquab, M., et al.: DINOv2: learning robust visual features without supervision. Trans. Mach. Learn. Res. (2024), https://openreview.net/forum?id=a68SUt6zFt, featured Certification
16. Ouyang, J., Adeli, E., Pohl, K.M., Zhao, Q., Zaharchuk, G.: Representation Disentanglement for multi-modal brain MRI analysis. In: Information Processing in Medical Imaging: 27th International Conference, IPMI 2021, Virtual Event, June 28–June 30, 2021, Proceedings, pp. 321–333. Springer-Verlag, Berlin, Heidelberg (2021). https://doi.org/10.1007/978-3-030-78191-0_25
17. Trivedi, A., et al.: Deep learning models for covid-19 chest x-ray classification: preventing shortcut learning using feature disentanglement. PLoS ONE **17**(10), e0274098 (2022). https://doi.org/10.1371/journal.pone.0274098. Oct
18. Woo, S., et al.: ConvNeXt V2: co-designing and scaling ConvNets with masked autoencoders. In: Proceedings of the IEEE/CVF Conference on Computer Vision and Pattern Recognition (CVPR), pp. 16133–16142, June 2023
19. Xia, P., et al.: Generalizing to unseen domains in diabetic retinopathy with disentangled representations. In: Linguraru, M.G., et al. (eds.) Medical Image Computing and Computer Assisted Intervention – MICCAI 2024, pp. 427–437. Springer, Cham (2024). https://doi.org/10.1007/978-3-031-72117-5_40
20. Xu, Z., Li, J., Yao, Q., Li, H., Zhou, S.K.: Fairness in medical image analysis and healthcare: a literature survey, October 2023. https://doi.org/10.36227/techrxiv.24324979, techRxiv preprint
21. Yang, Y., Zhang, H., Gichoya, J.W., Katabi, D., Ghassemi, M.: The limits of fair medical imaging ai in real-world generalization. Nat. Med. **30**(10), 2838–2848 (2024). https://doi.org/10.1038/s41591-024-03113-4, https://www.nature.com/articles/s41591-024-03113-4
22. Zhao, Z., et al.: Unobtrusive biometric data de-identification of fundus images using latent space disentanglement. Biomed. Opt. Express **14**(10), 5466 (2023). https://doi.org/10.1364/BOE.495438, https://opg.optica.org/abstract.cfm?URI=boe-14-10-5466

Author Index

A
Achara, Akshit 156
Adams, Lisa C. 125
Alleva, Eugenia 104
Alqarni, Maram 135
Alshamlan, Najd 32
Androutsos, Dimitrios 32
Anton, Esther Puyol 156
Aung, Nay 63

B
Bressem, Keno K. 125
Buser, Myrthe 125

C
Campanella, Gabriele 104
Cardoso, Jaime S. 208
Chang-Liao, You-Qi 22
Chen, Chi-Yu 1
Chen, Shengjia 104
Cheplygina, Veronika 115
Christodoulidis, Stergios 11
Chu, Yuseong 53
Clare, Kevin 104

D
Dang, Vien Ngoc 63
de Andrade, Paula Bresciani 177
De Neve, Wesley 198
de Paiva, Joselisa Peres Queiroz 177
Done, Susan J. 32
Dy, Amanda 32

F
Fay, Louisa 94
Feragen, Aasa 43
Fernandes, Leonor 208
Ferrante, Enzo 11

Forkert, Nils D. 74
Fournel, Joris 43
Fuchs, Thomas 104

G
Gatidis, Sergios 94
Gkontra, Polyxeni 63
Glocker, Ben 74, 145
Gonçalves, Tiago 208

H
Hammers, Alexander 156
Häntze, Hartmut 125
Hering, Alessa 125
Hu, Jingtong 167
Huang, Chin-Wei 1
Huang, Kuan-lin 104
Hudelot, Céline 11

J
Jegminat, Jannes 104
Jia, Zhenge 167

K
Khademi, April 32
Khaled, Zakaa 135
King, Andrew P. 84, 135, 156
Kori, Avinash 145
Kuo, Po-Chih 1, 22
Kushibar, Kaisar 63
Küstner, Thomas 94

L
Lee, Heejae 53
Lee, Tiarna 84, 135
Lekadir, Karim 63
Lin, Shih-Chih 1
Loureiro, Rafael Maffei 177

M
Matos, João 208
Mehta, Raghav 74
Mousavi, Seyed Amir 198

N
Nakayama, Luis 208

O
Oh, Byungho 53
Olegario, Tayran Mila Mendes 177
Osuala, Richard 63
Ozbulak, Utku 198

P
Paulo, Artur 177
Pedersen, Nikolette 115
Pegios, Paraskevas 43
Petersen, Eike 115, 187
Petersen, Steffen E. 63
Pinto, Bruna Garbes Gonçalves 177
Puyol-Antón, Esther 84, 135

Q
Qin, Xiaoli 32

R
Rai, Sweta 135
Rasal, Rajat 145
Rashid, Maariyah 135
Rashidian, Niki 198
Raumanns, Ralf 115
Reguigui, Hajer 94
Restrepo, David 11
Roschewitz, Mélanie 74
Ruijsink, Bram 84

S
Sankhe, Durva 135
Schumacher, Klaus 177
Sejer, Emilie Pi Fogtmann 43
Shafique, Abubakr 32
Shah, Partha 135
Shen, Mu-Yi 1
Shi, Miaojing 84
Shi, Yiyu 167
Shih, Kuan-Chang 1
Silva, Pedro Vinicius 177
Skorupko, Grzegorz 63
Sourget, Théo 11
Souza, Raissa 177
Stanley, Emma A. M. 74
Sutariya, Dishantkumar 187
Sydendal, Regitze 115
Szafranowska, Zuzanna 63

T
Tolsgaard, Martin 43
Tozzi, Francesca 198

V
Vakalopoulou, Maria 11
Vankerschaver, Joris 198
Veremis, Brandon 104
Verma, Ruchika 104

W
Willaert, Wouter 198
Wu, Yawen 167
Wulff, Andreas 115

X
Xu, Gelei 167

Y
Yang, Bin 94
Yang, Sejung 53

MIX
Papier aus verantwortungsvollen Quellen
Paper from responsible sources
FSC® C105338

If you have any concerns about our products,
you can contact us on
ProductSafety@springernature.com

In case Publisher is established outside the EU,
the EU authorized representative is:
**Springer Nature Customer Service Center GmbH
Europaplatz 3, 69115 Heidelberg, Germany**

Printed by Libri Plureos GmbH
in Hamburg, Germany